NO SURRENDER

A FATHER, A SON, AND AN
EXTRAORDINARY ACT OF HEROISM THAT
CONTINUES TO LIVE ON TODAY

CHRIS EDMONDS

AND DOUGLAS CENTURY

HarperOne
An Imprint of HarperCollins*Publishers*

HarperOne

Image credits: parchment frame on pages 1, 37, 105, 171, 231, and 291: SCOTTCHAN/ Shutterstock, Inc.; page i: ESB Professional/Shutterstock, Inc.; pages 1 (*left*), 8, 10, 12, 37 (*left, top, and right*), 49, 53, 58, 231 (*bottom*), 237, 238, 291 (*top left, top right, and bottom right*), 295, 314, 316: courtesy of the author; pages 1 (*top, bottom, and right*) and 105 (*bottom right*): Alamy Images; pages 37 (*bottom*), 83, 105 (*bottom left*), 171 (*right*), 188, 212: Roddie W. Edmonds Family; page 66: courtesy of Lester Tanner; page 71: James D. West, www.IndianaMilitary.org / contributed to Indiana Military website by Madalynn Mallo, daughter of William H. Benge, 1560th Signal Corp photographer; page 72: courtesy of Frank Cerenzia and Family; page 80: courtesy of Sydney S. Friedman Estate; page 95: courtesy of Sonny Fox; page 105 (*top right*): unknown German Kriegsberischter, Captured German Records Branch, US National Archives, courtesy of the Carl Wouters Collection; page 105 (*top left*) and 231 (*right*): courtesy of the Carl Wouters Collection; page 110: "The Siegfried Line Campaign" by Charles B. MacDonald, US Army Center of Military History; page 145: Wikimedia Commons; page 151: courtesy of Paul Stern; page 171 (*left*): photo by US Army Air Corps, courtesy of the Carl Wouters Collection; page 171 (*top*): photo by US Army Signal Corps, courtesy of the Carl Wouters Collection; page 171 (*bottom*): courtesy of the Jewish Foundation for the Righteous; page 231 (*left*): William C Allen/AP/ Shutterstock; page 231 (*top*): Trutzhain Memorial and Museum; page 275: courtesy of the Stern Family; page 291 (*bottom left*): Obama White House, provided by Roddie W. Edmonds Family; page 313: courtesy of the Embassy of Israel/Shmulik Almany.

HarperCollins books may be purchased for educational, business, or sales promotional use. For information, please email the Special Markets Department at SPsales@harpercollins.com.

FIRST EDITION

Designed by Leah Carlson-Stanisic

Library of Congress Cataloging-in-Publication Data has been applied for.

ISBN 978-0-06-290501-7
ISBN 978-0-06-299150-8 (ANZ)

19 20 21 22 23   LSC   10 9 8 7 6 5 4 3 2 1

To my Dad, Roddie Edmonds, and the other righteous among us.
May their tribe continue to increase, now and forever.

PART I

Look back and smile on perils past.

—SIR WALTER SCOTT

ONE

———

THREE HOURS BEFORE DAWN on Thanksgiving Day 2005, I startled from a peaceful sleep. I was sweating, murmuring the words of my late father, Roddie. I had been dreaming about his war journals, neatly written words, jotted in pencil, in his own cursive hand. "No one can realize the horrors the Infantry soldier goes through," he wrote. "You get scared and I mean scared, and don't let anyone tell you that he wasn't scared."

Dad never seemed scared of anything. When he was alive, he was always fearless, with the kind of quiet Christian faith that made him seem invincible. He always marched to the beat of faith, hope, and love. But never fear. Not once in all the years I knew him.

Lying awake, staring up at the dark ceiling, my mind flooded with confusion, and I whispered Dad's favorite Bible passage, Romans 8:37–39.

Yet in all these things we are more than conquerors through Him who loved us. For I am persuaded that neither death nor life, nor angels nor principalities nor powers, nor things present nor things to come, nor height nor depth, nor any other created thing, shall be able to separate us from the love of God which is in Christ Jesus our Lord.

What could have happened over there to make Dad, armed with such strength, so afraid?

For Dad, "over there" meant Europe during the Second World War. It meant Ziegenhain, an obscure town in the Rhineland-Palatinate of Germany, where he spent the final months of Nazi tyranny. He hit the beaches of France several months after the initial D-Day landings, slogged through freezing rain and mud in the fall of 1944, and saw horrific combat in the icy forests of Belgium that final brutal winter of the war, fighting along the supposedly impenetrable Nazi Siegfried Line in the Battle of the Bulge.

I knew little about Dad's service during World War II. Like a lot of members of the Greatest Generation, he never discussed the grisly details—I had only read about them in history books: dark skies exploding with relentless mortar fire; officers and infantrymen, already frozen in the winter cold, stunned by the ferocity of the barrage, a thunderous assault of 88 mm artillery fire followed by wave after wave of panzers and Nazi troops.

Now wide-awake, I sat on the edge of my bed for a few moments. It was a chilly morning in Maryville, Tennessee—the sun hadn't come up yet. I checked my cell phone. The temperature was slightly below freezing. I looked at my wife, Regina, who was still asleep beside me, just as she had slept every night for the past twenty-eight years. I pulled the comforter over her shoulder to protect her against the chill.

I rose from bed, stunned and confused by Dad's wartime confession, and stole away to the bathroom, where the sense of regret overwhelmed me. I splashed ice-cold water on my face to calm myself. Staring into the mirror, I saw a reflection of brokenness, the sound of my father's voice still ringing in my head.

Why didn't I ask Dad more questions before he died? Why did I let him take it to his grave?

I had been thinking about my father's wartime experiences the previous evening. Just after dinner, as I had helped Regina clear the table, our daughter Lauren had arrived home from Maryville College, where she and her identical twin, Kristen, were earning their degrees in education. Lauren was excited to tell us about a new history project: to interview a family member about a notable experience in their life—ideally an oral history of that person.

"Dad, when I told my group that Papaw was a POW in World War II, they said he was *definitely* the story, even though he's no longer alive. What do you think?"

I told Lauren I thought it was a great idea. "If I were you, I'd begin with his wartime diaries. Nana has them tucked away somewhere in her house."

"That's amazing," Lauren said. "Have you seen them?"

"I have—I've read them several times," I said. "You know, it's hard to imagine, but Dad never talked about them."

"Not even to Nana?"

"No, not even to Nana."

Later, while I was cleaning up after dinner with Regina, I stopped wiping down the countertop to reflect on Lauren's new excitement about her papaw Roddie. She was born in 1985, six months before he died. She knew little about him. But it seemed Dad had known her and Kristen instantly. "Two peas in a pod," he had said when they were born. "Can't tell 'em apart. And they're joining our third li'l pea, big sister, Alicia Marie."

Inseparable, they were typical sisters, my three girls, growing up in the '80s and '90s. They excelled at school, played house, had every Barbie and accessory created, loved Teddy Ruxpin, were cheerleaders, took gymnastics, and played softball, basketball, and volleyball. Friday nights they gathered to watch *Family Matters*, *Step by Step*, *Boy Meets World*, and *Full House*. Alicia, a future cosmetologist, loved to fix their hair, nails, and makeup, even if it was pretend. Papaw Roddie was right—they were three "peas in a pod."

Dad had always seemed to be right about people. Everyone acknowledged he had an uncanny ability to read a person's heart and instinctively see their true character. It wasn't judgment—more understanding, empathy, and helpfulness. This had served him well all his life. He used to spend his weekends visiting homeless shelters, churches, nursing homes, and VA groups in Knoxville just to sing and to encourage people in danger of losing hope—offering up his time, in *service*, simply because it was the right thing to do.

I wished my girls had known him. And Lauren deserved to learn as much as she could about her grandfather—especially his experiences during World War II. Of course, I knew the broad strokes of Dad's wartime experiences. I knew he had served as a master sergeant in the US Army. I knew he had fought with the 422nd Regiment of the 106th Infantry Division—the Golden Lions. I knew he had been a prisoner of war captured sometime during the Battle of the Bulge. But that's all I knew—not much, to be perfectly honest. And I was ashamed I didn't know more, that I hadn't asked him more questions when I had the chance.

I was in my mid-twenties, an unemployed father of three, when Dad passed away at the age of sixty-five. He died from congestive heart failure at his Knoxville home on Drifting Drive on August 8, 1985, twelve days from his sixty-sixth birthday. Dying at home on his own terms had been Dad's choice—it was his decision. Faith, family, friends, freedom—that's how he lived. And that's how he died too. Determined to enjoy his last few months at home, he had checked himself out of the hospital.

As I stood in the kitchen, I must have been a thousand miles away, lost in my own thoughts. Regina put her hand on my shoulder and asked me what was wrong.

"I feel like I'm letting Lauren—and her sisters—down. I should know more, but I just don't, and I can't even tell you why I don't. Had I not cared enough to ask him? I hardly know anything about Dad's childhood during the Depression or what he was like in high school. Or even his military service. It was like Dad lived an entire lifetime before I was born."

Regina, who had always loved my dad, listened, then reminded me that he had never really talked about his past. "He just lived in the present. That was Roddie. He was a wonderful dad and a wonderful person. A terrific grandfather too."

"He was," I said. "He was like a hero to me."

"Yeah, he was the same to me," Regina said.

"You know, I recall Dad saying in his diary, 'If a man lives through a major engagement, he isn't much good after that.' But he *was* good after that. I don't know all that he went through over there, but he didn't seem to have had any aftereffects from the war."

"I know," Regina said, "your dad was so high on life."

"Exactly. No matter what, he was always up. I remember him telling me, 'Son, life may knock you down, but you can always fly high as a kite.' That's what he did. He was a high flyer—a blessing and an inspiration."

"He sure was," Regina said.

Dad was always my hero—not because he'd survived the Depression and was a member of the Greatest Generation, an infantryman who'd fought in Europe to save the world from fascism. Nor because he'd remained in the army guard, and five years later, at the age of thirty, found himself on the front lines again, this time in Korea.

No, Dad was a hero in more ordinary ways: coaching my Little League team, teaching me and my buddies how to lay down a perfect bunt, how to read the opposing pitcher's windup, and when it was safe to steal second base.

He was a sincere Christian too. His love for God and other people was infectious, a model for living I still try to live up to. It didn't matter who they were. Dad cared about everyone he met. "Because God is good we must be good to others," he often said. Maybe that's why he always closed our family prayers with a simple request: "Lord, help us help others who can't help themselves."

During church fellowships—potluck gatherings after worship at West Haven Baptist Church—Dad would stroll through the room, flashing his warm smile, sharing his quick wit and genuine love for his church family. He had a way of lighting up the room and warming everyone's heart with fun and laughter. He was everyone's friend.

Dad also was a quick thinker, though he never attended college. I never asked why—as bright as he was—my dad didn't take advantage of the GI Bill.

He was well-read, artistic, and witty, and had an innate gift for wordplay. He could solve crossword puzzles almost instantly, often with only a cursory glance. A lot of time I would be sitting at the kitchen table, unable to get halfway through a tricky puzzle in the *Knoxville News Sentinel*. Dad would come in and look over my shoulder: "Son, don't you see the word 'alabaster' there? Then down from the *b* you can get 'bannister.'" In about three minutes, without even a pencil, he'd have completed the puzzle in his head.

Thinking about my childhood back in Knoxville, I couldn't help but grin right there in my kitchen, a soapy sponge still in my hand. My older brother, Mike, and some of our buddies from the neighborhood would assemble in our backyards and the nearby woods, reenacting battles, playing army. I used to dress up in Dad's US Army shirt, complete with the master sergeant's gold chevrons above three inverted "rockers" and his First Cavalry patch on the shoulder. The long-sleeve khaki uniform was so large that when I ran, the shirttail flapped down to my knees.

Some afternoons I would wear Dad's rectangular garrison cap, slinging the World War II canteen, filled from the garden hose, over my shoulder. Then we'd all chase each other through the sprawling neighborhood with toy rifles made from tree branches, pretending we were in the thick of the Battle of the Bulge in the winter of 1944. Those were good days.

My father—five foot five, stout and strong, with a thick head of wavy, sandy hair—would sometimes see us kids playing and smile as he mowed the front lawn. He loved having a Dutch Masters Corona De Luxe in the corner of his mouth when mowing and especially after dinnertime. He had picked up the habit during his military service, but after the war he rarely smoked them, just clenched them lightly between his teeth. He could make one soggy-tipped Dutch Master last for days.

"Son," he told me once, "these are my 'seegars'—'cause I like to 'see' how long I can make 'em last."

When I was fifteen, some buddies and I were asked to provide entertainment at our church's annual Valentine's Day banquet. This was a big deal, and we wanted to give the audience a performance they would remember forever. It was around the time when *American Graffiti* came out, and it brought a renewed popularity in '50s music, so my friends and I decided to perform some doo-wop classics. We called ourselves Big Daddy Clive and the Swingin' Five. I sang lead on the Earls' song "Remember Then." As my background singers bobbed and sang in four-part harmony, I jumped out from behind the choir rail wearing a pair of Dad's dressiest shoes—so oversized that I'd had to stuff the toes with newspaper just to keep them from sliding off my feet. Without missing a beat, Dad promptly jumped up from his front-row seat and pointed his finger at the stage. And in his booming baritone voice, which blended beautifully with the rhythm of the song, he yelled, "*There're* my white shoes! I wanted to wear them tonight and wondered where they ran off to!" The entire church erupted in laughter and cheers.

Dad had that kind of effect on people. Everybody loved Roddie. In his fifties, he had been a sales manager. His business was

selling prefab modular homes, and throughout Knoxville he was known for being scrupulously fair. He'd seal a deal with a handshake. He was honest, hardworking, and cared about his customers. And his customers cared about him. "I want to see Roddie," prospective buyers would say. Past customers would tell their friends who were looking for a modular home, "Just go see Roddie—he'll treat you right."

Roddie—I mean, Dad—was what other folks called a "square

shooter." Through that clenched Dutch Master, he loved to drop homespun Appalachian aphorisms, like "Boy, I'm as happy as a mule eatin' briars."

One Sunday, as he and I were sitting in the choir at church, the organist began the chords of Stuart K. Hine's gospel classic "How Great Thou Art," one of Dad's favorite songs—a song he sang so often as a solo it became his personal testimony. He whispered in my ear: "When you sing, Son, remember—sing by *letter*. Open up and let her fly." Dad could really belt out a spiritual tune with conviction:

And when I think that God, his son not sparing,
Sent him to die—I scarce can take it in:
That on the Cross, my burden gladly bearing,
He bled and died to take away my sin.

Dad's booming voice had been powerful, and his passion for God had been real and spiritually moving.

Back in the kitchen that evening, finally snapping out of my thoughts, I'd told Regina, "I was just thinking about Dad singing."

"You know your dad's voice could carry a country mile," Regina said.

"I just heard him singing. I know this sounds crazy, but it felt like he was right here."

LATER THAT NIGHT, we had gone to my mother's house a few miles away. I had told her about Lauren's class project and asked about Dad's diaries.

My mom, Mary Ann, disappeared into her bedroom. In a dresser drawer she had squirreled away important papers in one of Dad's old cigar boxes: Dad's death certificate, military discharge papers, expired passport, life insurance policy—official documents that were filled out with the details of Dad's life but never told the full story. Dad was fifteen years older than Mom when they married in 1953, so I only knew the story of the last twenty-seven years of his life. But there was much more to learn. There's always more to learn.

At the bottom of the cigar box were two fragile, nearly forgotten diaries from 1944 and 1945. They were the size of paperback

books. Narrow and light enough for an infantryman to tuck into his pocket. The cardboard covers were faded blue, and the pages inside were brittle and yellowed. Holding his diaries in my hands, I felt blessed—blessed to have been born to a loving father and mother, good people who worked hard, loved their family, helped their neighbors, served their community, and honored

God. But I felt burdened too. Life's blessings can sometimes add a weight of responsibility. *To whom much is given, much is required.* I felt a growing responsibility to myself and my family to know what had happened to Dad.

Back home, Lauren scanned the diaries, combing through the tender pages for details to use in her presentation. Regina and I listened as she read aloud from her papaw's fading penciled script: "'A lot of things I am not going to write,'" she said, fighting back tears. "'Because they aren't exactly nice to talk about. I know God was with us, and he answered our prayers. I learned men, even better than before. Some were good, some were bad, some were better, and some were worse.'"

Watching her struggle to get through the diary entry, Regina and I both started tearing up. I realized this was the closest Lauren was ever going to get to her papaw, reading a few lines he wrote out in the winter of 1945, sixty years earlier—twelve years before I was born, decades before Lauren came into our lives.

Continuing, she traced her finger on another line. "'I will not tell anyone the true happenings—by true, I mean some of the worst things that happened to us as POWs.'" Then she turned to me. "Did Papaw ever describe his experiences to you?"

"No, I asked Dad about it when I was your age. All he said was 'Son, we were humiliated.' I asked a bunch of times, but that's all he'd ever say. That one word, 'humiliated.'"

Lauren shook her head and said, "I can't imagine what Papaw had to go through." Then she closed the diary with a kiss and set it down gently on the dining room table.

THE ICE-COLD WATER stung my face. I turned off the bathroom light and went to the dining room to read the diaries. Other than the ticking of our Coca-Cola clock as it hit 3 a.m., the house was dead quiet. The floor beneath my bare feet was cold, colder than usual. Sitting at the dining room table, I flipped through Dad's notebooks for the first time in decades, which kind of felt as if I were walking through an old house, with corridor after corridor leading to new secrets. In those wartime diaries I could *hear* Dad's voice, alive and vivid, as if he were sitting at the table with me. But Dad and his past were still distant and unknown. Each word was precious, every phrase meaningful, but his powerfully chosen words weren't enough to unlock his past.

I had read the diaries when I was in college, but now, after all these years, there I was again, a middle-aged man, a father, an ordained minister, still trying to make sense of my dad's words from when he was twenty-five years old. But the spark of Lauren's school project had lit a fire inside me. I couldn't put the diaries away. Sitting in shadows, with only a single light above me, I felt the diaries come to life. As I read through them again and again, Dad's mysterious words started to hit me with new meaning. I had forgotten, but there were even illustrations, floor plans, and menus for a restaurant called The Jolly Chef that the starving POWs probably fantasized about opening after the war. A talented graphic artist, Dad had designed the logo and hand-lettered the menu. I wondered: Could it be that he and three other would-be restaurateurs—along with their captive customers—were eating off an imaginary menu?

There were also several pages of names and addresses, some of the men who had survived the camp with Dad:

Arthur Levitt—Bronx, N.Y.

Emanuel Frawert—Paterson, New Jersey

Morris Chester—Baltimore, Maryland

Sydney S. Friedman—Shaker Heights, Ohio

John V. Selg—Brooklyn, N.Y.

Carl E. Johnson—Grand Rapids, Michigan

Ward R. Richardson—Ludington, Michigan

Edward W. Berry—Grand Rapids, Michigan

Ralph H. Wahlman—Willisville, Illinois

Carl Falch—Miamiville, Ohio

Henry E. Freedman—Dorchester, Mass.

Paul Stern—Bronx, N.Y.

These boys—and they *were* boys, eighteen and nineteen years old, not even as old then as Lauren was when they were taking on the Nazis—had come from every part of the country, young soldiers from different backgrounds and cultures fighting together as one for a common cause. Their names, seemingly written in their own hands, added to my sense of mystery.

Dad had written briefly about the ferocious battle, the courage of his "boys," a few details about their capture and the harsh conditions at the Nazi camps. His descriptions were terse. Bare facts. Sometimes just fragmented sentences. Mental notes. Personal shorthand. Words clearly scribbled in haste.

Dad's war had ended several decades ago. I'm sure he had seen his fair share of horrors, more than enough to last a lifetime: young comrades killed and wounded during battle. How had he coped with the humiliation of being captured, of not fighting to the death? How deep had the guilt been, knowing he had survived the fighting but, perhaps, should have died with his buddies?

And yet I couldn't escape the fact that, despite the horrors hinted at in their pages, Dad's diaries had been written with fierce faith, an unwavering belief in God and a higher purpose, something bigger, and a greater promise on the horizon, full of light, life, and love.

Lord, I don't know what happened over there, but you do.

Please help me find out.

TWO

———

L AUREN AND HER CLASSMATES earned an A on the project. But before their grades had been posted, I knew my amateur sleuthing had only just begun.

I was in my pajamas, sitting at my Dell computer in our spare bedroom, which doubled as a home office, while Regina slept down the hall in our bedroom. It was just past midnight on a February night in 2009 with a light freezing fog outside frosting the windows. I double-clicked the Google icon and typed: "Master Sergeant Roddie Edmonds."

I expected to be directed to a World War II national archive database or the 106th Infantry Division veterans page, but the first link that popped up was an article in the *New York Times* from July 30, 2008, titled "Richard Nixon's Search for a New York Home," which recounted how the disgraced former president had tried to buy an apartment in Manhattan. No one would sell to him following his humiliating resignation from the White House. No one wanted Mr. Nixon to be their neighbor. Then a prominent Harvard-trained attorney stepped up. Lester Tanner was a lifelong Democrat. He'd been friends with Bobby Kennedy and acted as a delegate for him during his 1968 presidential campaign prior to Kennedy's assassination. Although Mr. Tanner disagreed with Nixon's politics, he told the reporter, he was nonetheless appalled by the blackballing. So he decided the moral thing—the *right* thing—to do was to sell Nixon his place, a 12-room town house in the middle of Manhattan.

The Tanners had never told the story before, and I read it trying to figure out what it had to do with Dad.

Questions rushed through my mind.

The New York Times?

Nixon?

Kennedy?

Harvard?

Multimillion-dollar Manhattan town houses?

None of it made any sense.

But then, almost as a side note in the article, Mr. Tanner mentioned that as a GI in World War II he had been taken prisoner and held in a Nazi stalag near Ziegenhain, Germany. He went on to say that the "brave officer," Master Sergeant Roddie Edmonds, had defied the camp's commander, saving Lester's life.

I was stunned.

I checked Dad's rank, "master sergeant." It was correct. The spelling of his name was right too. It had to be Dad.

I sprang from my chair and ran to the bedroom. When I flipped on the lights, I nearly scared Regina out of her wits.

"Wake up, wake up!" I yelled, trying to rouse her out of bed. "You've got to see this—it's unbelievable—"

"Can't this wait till morning?"

"No, it can't—please. Get up. Hurry! You've got to see this. Come on."

Regina stumbled to the computer and started reading the article. I stood behind her, waiting until she got to the paragraph about Dad. A part of me still couldn't believe it, and I was hoping Regina wouldn't tell me I had misread it or that Mr. Tanner couldn't possibly be talking about Dad.

"I can't believe it," Regina finally said. "It's got to be Roddie."

Then we were ping-ponging questions.

"Chris, who is Lester Tanner?"

"I have no idea, but I would love to know. Who *is* Lester Tanner?"

"It says he's a Manhattan attorney—wonder if he's still practicing law."

"Wouldn't he be in his late eighties by now?"

"The article's almost a year old."

"You think he's still alive?"

"What did he mean when he said your dad defied the camp commander?"

"I don't know, but it sure sounds like Dad saved Mr. Tanner's life."

After reading the story for about the fifteenth time, I finally scrolled down to the comment section. Several readers had already written messages about President Nixon's difficult return to New York. I don't know what I was hoping to find there, maybe some more information about my dad. Or maybe some recognition from readers about what he did. It was mostly notes about politics, how Democrats and Republicans needed to get along, and New York real estate, how Nixon probably wouldn't even be able to afford to live in the city today.

I decided to add my own comment—a simple note, addressed

neither to the story's writer, Ralph Blumenthal, nor to Lester Tanner. I wrote it to Dad.

> Thank you, Dad, for doing what was right and for living out your Christian values. Your heroic actions continue to touch our lives. A proud son. Chris Edmonds

For the next couple of weeks, my questions continued and my curiosity increased.

I dug into Dad's diaries, looking for something that might give me a clue, possibly a phrase, a sentence, or even a name of a POW—anything that would start to help me answer even a few of the questions piling up in my head. *Who was Lester Tanner? What was Dad's act of bravery? Why had he never mentioned it?*

Most of Dad's entries were clear and specific, but others made little sense. Reading them all was both exhilarating and frustrating: the more I studied the diary, the less I seemed to know—a puzzle. This time, though, I didn't have Dad looking over my shoulder to help me make sense of his entries. What was really driving me crazy were the missing pages. From the slightly raised staples and the worn fray of the remaining pages, it looked as if Dad had ripped out a few pages, right in the middle of things. Dad's story was interrupted abruptly in midsentence.

I enjoyed my last meal on the evening of the 17th, because the morning of

---

And then nothing. Just blank pages. No finished thought. No indentations pressed into the blank pages from a heavy pencil. Nothing but more mystery and even more questions.

Where were the missing pages? Lost? Still hidden in Mom's dresser drawer? Or had Dad intentionally removed them or possibly destroyed them for some reason?

On the page following the blank ones, the diary continued with no connection to his previous thoughts.

I guess the reason I am writing this, mainly, is to relieve my mind, and while some of the events are fresh in my mind.

Or maybe the mystery lay in the pages he hadn't touched, the empty pages he'd left without a mark. Why would he have left so many empty pages in the middle, then begun writing again at the back of his journal?

Had he done it on purpose? Had it been to fool the Germans if he ever had to show the notebook to them? Or had things become so bad that he hadn't had the will or strength to write?

Perhaps the mystery was hidden in these cryptic notes scribbled in haste on the last two pages:

Army career

Capture

First night and the long march

Gerolstein

Trip to Bad Orb

Jewish moved out

Dogs

Hiding

Or maybe in another mysterious phrase that ended abruptly:

Before the commander

If only Dad were here to explain, I thought. If only I had been more interested in Dad's life and less in my own—if I had just spent more time with him maybe . . . maybe . . .

Then I saw it. Right there, a few pages into the list of POWs.

**Lester J. Tannenbaum
1384 Grand Concourse
Bronx, N.Y.**

Lester Tannenbaum. Yes, that *had* to be Mr. Tanner, despite the abbreviated name. He had been in the diary this entire time, along with the address of where he had lived before the war—which appeared to be written in Lester's own handwriting. I wondered if any of the other men listed with Lester in the diary—Henry E. Freedman, Arthur Levitt, Emanuel Frawert, Morris Chester, Sydney Friedman, Paul Stern—were still alive and had their own stories to share that could help me learn more about Dad's.

WEEKS FLEW BY. Then weeks turned into months. Before I knew it, three years passed as I got caught up in the frenzy of life. I changed careers. Kristen and Lauren graduated college, then both got married within a few months of each other. Soon both had kids of their own—giving Regina and me our second and third grandchildren.

My new job as executive director of a nonprofit consumed me. The organization mentored middle school students on twenty-seven campuses spread across five different counties in East Tennessee. I had volunteered with YOKE Youth Ministries since 1978, when I was still in college, and YOKE was like family. I had worked in some capacity, from middle school mentor to board member, for pretty much my entire adult life. Now I was its director. I loved the work, the YOKE family, and middle school kids. I was also serving as pastor of Piney Grove Baptist Church in Maryville. Life was a blur. It certainly wasn't dull, I'll tell you that. Every now and then I tried to track down Mr. Tanner, but time passed quickly.

One night, though, I was watching the news with Regina. A report mentioned that nearly three hundred veterans of World War II were dying each day.

I was shocked.

With each passing day the frustration of not knowing what had happened in Germany grew inside me, along with my desire to find Mr. Tanner, the only man who could possibly help me unravel this mystery.

But I didn't know if he was still alive. I started worrying that I had missed out on my chance to find him. I prayed aloud: "Lord, your timing is always perfect. Show me the way I should go. Please let Mr. Tanner be alive and help me find him."

Finding Lester became a priority. I searched the internet and found some leads. At the same time, Lester had started looking for me. Because of the comment I had left on the *Times* website, Lester and his daughter, Shari, had begun their own search. Shari had even enlisted some of her friends. They eventually found my email address on the YOKE website, and Lester sent me an email.

My prayers were answered. Not only did God send Lester my

way; he also sent me a short letter of introduction and Lester's home phone number.

I immediately picked up the phone and dialed. I was excited, grateful—but also nervous. The phone rang and rang.

What should I say?

Finally, I got an answer—but it was Lester's voice mail. I left a long message and promised to call back.

The next few days I was busy preparing to take more than two hundred middle school kids and adult mentors to a weekend camp. Camp Ba Yo Ca was an idyllic Christian retreat nestled in the foothills of the Great Smoky Mountains. It was a special place. In fact, I had met alone with God there many times to rekindle my spirit, make big decisions, or simply rest in his beautiful creation, including the mountains Dad had loved so much—mountains of blue and smoky gray grandeur. Dad had grown up in South Knoxville just blocks from Chapman Highway, the "Gateway to the Smokies." For Dad the Smokies had been the gateway to his heart, his home, and his heavenly Father.

I loved the Smokies too—but not as much this weekend because I would be isolated with no cell service or internet. I was hoping Mr. Tanner would call before I hit the dead zone in Happy Valley. But no call came. Mr. Tanner would have to wait.

On Sunday, as the kids headed home, the camp grew quiet. After helping clean up and load supplies, I drove past the camp sign, through the creek, up winding Happy Hollow Road. I wanted to get over the mountain to the spot all camp leaders knew well—the spot where you could get a cell signal.

Anxious, I checked my voice mail. Sure enough, Lester Tanner had called and left a message. I listened and relistened to that voice—deep, with a dignified New York accent, and remarkably robust for a man of eighty-eight.

A few days later, as Hanukkah and Christmas approached, I

called Manhattan again and heard a "Hello." I was thrilled: one man from the Greatest Generation and one from a *grateful* generation were finally connected, mysteriously bonded by something that had happened nearly seventy years ago. To Mr. Tanner, it was as vivid as yesterday.

Listening to him, I felt a sense of awe, thinking of the small miracles that had led to this moment. At first, the conversation was mostly about our work, our families, and our grandchildren. I could tell that it was hard for Mr. Tanner to hear on the phone. With Mr. Tanner's prompting, I agreed to email him rather than continue our conversation on the phone.

While Mr. Tanner had a difficult time hearing me, I heard Mr. Tanner loud and clear. He said he was awed by Dad's bravery and eternally grateful for actions that helped save his life—*and* the lives of others.

"Dad saved *other* men too?" I said under my breath, mostly to myself, because I sensed that my journey was only just beginning.

THREE

O<small>THER THAN DAD</small>, no one in my family had ever been to New York City. For whatever reason, I just had never felt the desire to visit. Now I couldn't wait. Neither could Regina nor my ten-year-old grandson, Austin. Here we were, seemingly overnight, preparing for a trip to Manhattan—the Big Apple—to meet a soldier from my father's generation named Lester Tanner. A respected attorney who might tell me things about my father I had never known. A person who might help me unravel the mystery of Dad's diary and his time as a POW.

"My wife, Regina, doesn't like to fly," I had told Lester when we were finalizing plans, "so we're going to drive. I've been researching places online, but honestly, I'm out of my depth. I don't know anything about New York, and I'll need a place to park the car in Manhattan."

"Might I suggest the Harvard Club?"

"Can I get a room at the Harvard Club?"

After a long pause, Lester said, "No, you can't. But I can."

"How much is that going to cost per night?" I asked, more than a little nervous about the rate.

"For you," Lester said, "nothing. Chris, it's the least I can do for the son of the man who saved my life. I owe *everything* to your father."

Regina and I were stunned. We were headed to New York, staying at a private club just two blocks from Times Square. All because of the generosity of Lester Tanner, a man I barely knew. All because of something my dad had done for him nearly seventy years ago.

"Chris," Lester told me, "you should go on the website and make sure you're aware of club rules."

I logged on and immediately saw that the club had a *strict* dress code. I read it aloud to Regina, taken aback by its formal rules. "Says here that the 'accepted standard of casual business attire for men is collared shirts without jackets and ties' and 'for women, an equivalent standard of casual attire is required.' In the Main Dining Room, the standard is jacket and tie for supper, and women have to wear dresses or suits of 'comparable formality.'"

I turned to Regina and tossed out a famous old hillbilly line: "Golly, babe, we're headin' to New York City—guess we're gonna need some *shoes*!"

But Regina wasn't laughing. She went right to the mall with our grandson Austin and his mother, Alicia, to buy him a suit.

"He's a ten-year-old boy living in East Tennessee," I said. "Why on earth would he ever need a dinner jacket, tie, and dress shoes?"

First thing Sunday morning, I packed up the family SUV, making sure to hang Austin's brand-new blue suit and crisp white button-down shirt in the back so it wouldn't get wrinkled. After I preached to my congregation at church, Regina, Austin, and I were on the road by the afternoon, hitting the highway north.

Though there were snow flurries in the forecast, an unexpected storm hit the East Coast, and I could hardly see the road ahead in what seemed to me a blizzard. We made an unplanned stop to spend the night at a motel in Winchester, Virginia. The next morning, in the nasty weather, I drove up I-81 toward New York City, the wind howling, rattling the SUV. To calm my nerves, I pulled into a gas station in New Jersey. When I headed for the pump, a huge guy with a *Sopranos*-like accent yelled at me:

"Hey! Whattaya doin'?"

"I'm just gettin' some gas."

"Get back in the car. Wait ya turn!"

I quickly slipped back into the driver's seat, which is when I saw the sign that said, "It is unlawful to serve yourself in New Jersey."

Austin asked, "What are you doing, Pop?"

"Waiting on our gas to be pumped."

"Why?" Regina asked.

"Because that big sign, and that big guy, said to wait."

And wait we did. When I finally merged back onto I-81 thirty minutes later, I turned to Regina and said, "Remind me to get gas *before* we get to New Jersey."

WHEN WE PASSED the exit for Hoboken, Regina got excited. She wanted me to stop so she could visit Carlo's Bakery and maybe meet Buddy Valastro, the star of *Cake Boss*, one of her favorite shows. Trying to tell her that we didn't have the time, I nearly missed the lane for New York, and it seemed that every *single* bus on the East Coast was trying to force its way past me to get into the Lincoln Tunnel.

Both Regina and I were overwhelmed by it all. We explained to Austin that we were now driving, in stop-and-go traffic, *under* a major river called the Hudson. To top off all the craziness, down in the Lincoln Tunnel I lost GPS service. I was driving way too slow for city traffic. I was getting honked at, cursed at, then I emerged from the tunnel in the *wrong* lane.

I lowered my window and asked an NYPD officer on foot patrol, "Officer, I need to turn left. Can you get me over?"

To which the annoyed cop shrugged and yelled, "Wrong lane, buddy, keep right!"

We spilled out onto the streets of Manhattan, utterly lost and in a bit of a panic. Somehow I was headed downtown. And worse, the rain mixed with snow made driving even more difficult. As the snow floated down, flying upward were our prayers. Getting to the Harvard Club *soon* was the only thing on my mind.

The tall buildings and winter weather kept interfering with the GPS. Frustrated with the traffic, flashing neon lights, and jaywalking, I wasn't making much progress on the one-way streets of Midtown. After several minutes, I started to recall the map of Manhattan I had pored over in preparation for our trip. I tried to remember the turns I needed to make. It wasn't easy, but my recall and survival instincts finally kicked in. I started to be more aggressive, trying to drive like a real New Yorker. I adjusted my country style to the big city. I swerved and accelerated into open lanes, I squeezed in, rode bumpers, and honked my horn, the entire time Regina telling me, "You're going to get us killed!"

In short order, we found ourselves on West 44th Street, in front of the Harvard Club. I double-parked the SUV to unload the luggage, just as Lester had directed. I didn't even blink when a long line of angry drivers honked to get past me, because we had arrived safe and sound, in grand New York style—a beautiful thing.

"I *love* the city!" I told Regina, who just shook her head.

Inside, the buttoned-up clerk behind the front desk heard our thick Tennessee accents, gave us the once-over, as if Mister and Missus Gomer Pyle had mistaken the Harvard Club for a Best Western. "Are you absolutely *sure* you're at the right place?" the clerk asked.

"Yes," I assured him. "We are here as guests of Mr. Lester Tanner. He made the reservations for us."

The clerk changed his spirit, and the check-in went well. As we made our way to the elevators, we were gob-smacked: the floors of the Harvard Club were covered in plush crimson carpet, the walls in high-gloss crimson paint.

After settling into our room, I guided Austin on a tour of the halls, which were lined with portraits of Harvard graduates—a Who's Who of US history: Teddy Roosevelt, Franklin Roosevelt, John F. Kennedy. There were framed portraits of Supreme Court justices and playwrights and novelists, Nobel Prize laureates and movie stars and Civil War heroes. I paused under an immense portrait of President Obama. "Austin, did you know the president is a Harvard Law School graduate?"

Hanging from the walls was a collection of immense swordfish and sailfish, the heads of moose and antelope and, in the cavernous main hall, a majestic Asian elephant with three-foot-long tusks, shot and stuffed during the reign of Queen Victoria, which stood out as a landmark of taxidermy.

Later, in the dining room, dressed in our finest, we stared at the silver cutlery, at the crimson leather menus, which didn't list any prices. At least I was prepared for this surprise. Before our trip, Mr. Tanner had explained that no cash could be used at the club—and we listened as the waiter rattled off the specials of the day.

"Sir, do you have any questions?"

"No," I said. "Thank you very much."

We decided on our meals, then stared at each other, wondering when the waiter would return.

Meanwhile, the waiter was across the room—staring at us!

After about fifteen minutes, the waiter finally came over to our table.

"Sir," he said, "are you planning on placing an order?"

I told him I was.

We waited again in an awkward silence.

I finally said, "My wife would like—"

The waiter stopped me midsentence. "Sir," he said, "please write your orders down."

Only then did I understand that the small notepad with the stubby yellow golf pencil—the same one Austin had been doodling on—was how we were supposed to place our orders.

I glanced at Regina. We both laughed. We had an unspoken understanding that we were out of our element—having driven all the way from Tennessee to meet a man we didn't know in an overwhelming city neither one of us had ever expected to visit—and we would just go with the flow and laugh at ourselves along the way.

Before we ate, I offered one of Dad's favorite blessings: "God's eye is on the sparrow, and I know he watches me and you."

After dinner, we changed back into our jeans and stepped out to explore. Because of Mr. Tanner's busy schedule, we had to wait until Wednesday to meet him, so we would spend our free time seeing the sights. We walked toward Times Square. It was late, about 10 o'clock. Lots of people were out, but traffic was light. With few cars and a light snow falling, magic had descended on this winter night. It was cold *and* warm all at the same time and somewhat surreal.

"Just like the movies," Regina said, "a winter wonderland."

"Yes, it is," I said. "Just the way God planned it."

Austin laughed as he caught snowflakes on his tongue.

It was a terrific evening wandering through Times Square, enjoying the crowds, taking selfies in front of the famous lights, and visiting the iconic shops.

Back in our room, snug in our pajamas, Regina told Austin, "I'm sorry you had to see that, Bubba."

"Yeah, me too," Austin said.

"See what?" I asked.

"The painted lady with no top—she was naked from the waist up," Regina said.

"Yeah, Pop," Austin said, "it looked like she had a top on—but didn't."

"How did I miss a naked lady? Am I such an old man that I didn't see her?"

Regina nodded and said, "Yes, Chris, I guess you are."

The next morning, we got up early and took in as many sights as possible. We started off with the Empire State Building, where Austin took off running when a statue of King Kong came to life and tried to grab him—it was actually just a guy dressed up in a realistic gorilla suit. We shopped at Macy's . . . hit Broadway to get tickets for *Newsies* . . . visited Rockefeller Center . . . and walked up to Central Park—on the way crossing paths with some weird circus-like character who, in spite of the chilly temperatures, was bare chested and had a giant boa constrictor draped around his neck. We generally acted precisely like you'd expect three wide-eyed tourists from the hills of East Tennessee in New York City for the first time in their lives to act. We snapped pictures everywhere—even on the subway and in a yellow cab. Regina and I were especially fascinated by two street-savvy boys, no older than Austin, who skateboarded through heavy foot traffic and right onto the subway train.

Finally, on Wednesday, March 20, I eagerly waited for Lester in the lobby of the club. Every time an older man passed through the glass doors—stooped eighty-year-olds with horn-rimmed glasses and canes, tiny men who must have attended Harvard when Franklin Roosevelt was president—I thought it was Lester.

Suddenly, with a gust of brisk March wind, in walked a powerfully built older man of six feet, with a head of thick, swept-back warm-brown hair, wearing an elegant blue business suit, a starched white shirt, and a yellow silk tie with crimson ovals in a perfect Windsor knot.

I was speechless. This was not what I had expected. Glancing at the floor, I adjusted my jacket and straightened the knot in my own tie. I felt my face flush, and despite the cold breeze, my forehead beaded with sweat.

"You must be Chris," Lester said. "Yes, you do resemble your father."

I remember thinking, *How does he remember how my father looked?* It had been seventy years since they'd seen each other.

We shook hands warmly. I was surprised by the strength of Lester's grip; of course, he *must* have been strong to survive whatever horrors he and Dad went through together.

I tried to imagine how Lester must have looked in 1944, a young soldier serving alongside my father. Lester alone knew things about him that I could not possibly imagine. Meeting Lester, I realized, would be like meeting Dad as a young man for the first time.

Lester, like Dad, seemed so well-adjusted. It struck me that he and my father, somehow, overcame the horrors of war and came back home to live ordinary lives of extraordinary influence. For a moment I paused and held Lester's hand, in awe of him and Dad.

"Before lunch, we'll go up to the library," Lester said. "We can speak in private. And it will be much easier for me to hear than in the noisy dining room."

Even in the quiet of the library, seated at two small study desks, under the unblinking gaze of that famous stuffed elephant, Lester said he was having difficulty hearing. He scooted his chair right next to me, leaning in so close I could feel Lester's warm breath against my cheek.

"Your father was cut from a different cloth than the rest of us." His words were measured, precise, eloquent—as you'd expect from an experienced Harvard-educated litigator. But they were also warm, like he had known me for years and was talking to a long-lost friend. "Roddie was regular army, from the peacetime Old Guard, who found himself training young men to defend their country. His leadership style embraced the Infantry School motto: 'Follow me.' That's the shoulder patch of the Fort Benning command. He showed no arrogance or disrespect for his subordinates—whether they were junior noncoms [noncommissioned officers, also known as NCOs] or the recruits he was assigned to train. His style was simply to make us proud of what we were being called upon to do in defense of the republic. We knew he respected us. We knew that he was devoted to our survival in combat. But he never left any doubt that he was in command. He expected every one of his orders to be followed."

I shook my head in amazement. "Growing up, my dad was such a soft-spoken, fun-loving guy," I said. "It's hard for me to imagine Dad ordering other men around . . ."

"Your father led by example. He taught us well: 'Love your rifle; keep it clean at all times. Develop your skills on the firing range. Fix bayonets to defend yourself at close quarters.' So many of our regiment were green youngsters—barely out of high school. Boys who'd never seen a rifle in their lives, let alone fired one in anger. Roddie's experience and knowledge gave us all confidence. Your father was of a different generation, Chris. We called him 'The Old Man.'"

"'The Old Man'? Lester, my dad was barely *four* years older than you!"

"Four years?" Lester's eyes widened, and for a moment it looked like he might tumble out of his wine-colored leather chair. "Four years? I knew him so well, yet we never talked about our ages. My image of him as a master sergeant was that he had to have been in the army for many, many years . . . and that would've made him about thirty years old. His knowledge and conduct as the chief enlisted man in a regiment of five thousand soldiers enhanced that image of maturity. It was clear that he had been in the peacetime army and a high achiever for many years—long before I met him. Only *four* years older than me? Chris, that's astonishing."

There was a long silence: the only sounds in the library were those of other Harvard Club members in leather armchairs, flipping pages of the *New York Times* and *The Wall Street Journal*.

"Lester, I know it must be hard for you to talk about," I said, "but what exactly happened in that German POW camp? What did my dad do at Ziegenhain?"

"What your father did at Ziegenhain is one of the most remarkable things I've ever witnessed—it is the defining moment of my life. And, Chris, I should tell you, I believe your father is deserving of the Congressional Medal of Honor."

I was speechless. "The Medal of Honor?"

"The Medal of Honor." Lester nodded. "But before I tell you what happened in Germany—"

"Yes?"

"Please tell me more about your father."

PART II

*Leadership is intangible, and therefore
no weapon ever designed can replace it.*

—GENERAL OMAR BRADLEY

FOUR

M Y FATHER WAS BORN in late summer 1919, when his
hometown of Knoxville was about to erupt in turmoil.
Wednesday, August 20, began bright and unseasonably warm.
A severe drought had left the Tennessee River riding low as it
slipped south under the Gay Street Bridge. That night, at twenty
minutes past eight, Dr. J. J. Eller delivered Roderick Waring
(Roddie) Edmonds—the last of four boys—to my grandparents,
Thomas and Jennie, at their shotgun-style home on West Fifth
Avenue.

Ten days later, on August 30, a lynch mob stormed the county
jail in search of an African American former deputy sheriff named
Maurice Mays, who'd been accused of murdering a twenty-seven-
year-old white woman named Bertie Lindsey. Unable to find
Mays, the rioters looted the jail, released the white inmates, then
fought a pitched gun battle with the residents of a predominantly
black neighborhood just blocks from my grandparents' home. I'm
sure they could hear the shots firing outside their windows.

The governor ordered the Tennessee National Guard to disperse the rioters, which they did by firing their machine guns indiscriminately. Seven people were killed—six black, one white—and scores were injured on that bloody Labor Day weekend. Knoxville's race riot was part of a nationwide epidemic of ugly civil unrest known as the Red Summer. From May to October, violent racial incidents swept cities across the United States, resulting in an estimated six hundred deaths. The 1919 riot is still remembered as one of the worst episodes in local history, shattering Knoxville's image as a sleepy Southern city with a reputation for tolerance.

In spite of this, or maybe because of it, Roddie grew up in a home that rejected all forms of intolerance—especially the ignorance of racial hatred. His code was to follow the Golden Rule: love everyone. The foundation of Roddie's values was faith in Christ; his moral compass was the Bible. Even back then, I'm proud to tell you, the Edmonds were committed to deep Christian faith, which Roddie fully embraced as a teenager.

My grandfather Thomas Calvin Edmonds, or T.C., and my grandmother Jennie Mary Sexton Edmonds were born-again Methodists. Their relationship with God was practical, real, and sometimes raw—a singular devotion that affected every aspect of their lives. They didn't have many possessions, but they had what they believed mattered most: gratitude for life, respect for others, and an unwavering devotion to God. T.C. and Jennie lived by the biblical instruction to "train up a child in the way he should go, and when he is old he will not depart from it." They believed faith was as much caught as taught.

T.C. was a highly skilled wallpaper hanger who marked favorite verses, like Revelations 22:1–4, in his worn-out Bible. He had grown up singing in the gospel tradition known as the Old Harp way, which is probably where my dad got his talent for singing. Old Harp singing schools flourished in Appalachia, especially in East

Tennessee, where students first learned how to sing the "shapes"—the scale of *do-re-mi*—then learned how to sing the words. Like my father, T.C. had a powerful voice—singing both baritone and tenor—and was popular in the school's singing events, which drew hundreds of attendees from all over the state. Tagging along with T.C. to these events, Roddie came to love singing in the Old Harp style.

As far back as I can tell, the Edmonds family on both the paternal and maternal sides were known as "God-fearing folk" who believed that "the righteous shall live by faith." Etched on many of their tombstones were scriptures that reflected their lives, like Revelation 14:13, "Blessed are the dead who die in the Lord," or a simple statement of faith, "Asleep in Jesus."

Saturday, June 24, 1922, was unusually hot in Knoxville. Keeping up with four active boys between age two and a half and sixteen while T.C. was out hanging wallpaper for Wilhoit and Hayes Company left the newly pregnant Jennie exhausted. Robert Seymour, David Leon, Thomas Carlyle, and little Roderick Waring were well-behaved boys but full of energy—especially Roddie.

Had he lived, Richard Eugene, T.C. and Jennie's first son, would have been seventeen and could have helped her care for his brothers. But Richard had died suddenly, at just two years old, in August of 1904 from gastroenteritis. The tragedy of Richard's death lingered nearly eighteen years later. Jennie and T.C. expected this summer to be brighter, what with a new baby on the way who could be a playmate for Roddie.

Jennie was petite and captivating. Her blue eyes sparkled under her red hair, which was full of curls and finger waves, worn in the style of Ziegfeld Follies showgirl Gilda Gray. Though Jennie cherished being a mother, she had decided that, at age thirty-nine, this child would be her last. T.C.'s own mother, Mary, had encouraged Jennie to make Roddie her last, but Jennie wanted to try one more time, hoping the Lord would give her and T.C. a daughter to round out the Edmonds clan.

As they said their evening prayers and tucked Roddie into bed, T.C. and Jennie felt grateful—blessed. But Jennie also felt worn down. For the past few days she had been hoarse; her throat was tight, which made swallowing difficult. That morning she had noticed her neck was badly swollen, and her cough, which had started as a minor annoyance, had grown more painful and frequent throughout the day. Her breathing had grown labored. At first, the symptoms had seemed minor to Jennie—probably commonplace East Tennessee allergies or a summer cold. She didn't even mention anything to T.C.

But that night, at 2:15, she died, quickly and unexpectedly, tragically. Alive one minute, gone the next, along with the unborn baby. Both taken due to complications of a goiter, an enlargement of the thyroid gland caused by an iodine deficiency.

There was nothing T.C. or the physician could have done: the goiter had closed off her airway. Jennie had died from asphyxiation.

She was buried on Wednesday, June 28, 1922, in Woodlawn Cemetery in her beloved South Knoxville. T.C., who never remarried, recorded her death on the front and back leaves of his pocket-size Bible, which he carried with him the rest of his life.

THE DEATH OF Jennie left T.C. to raise the boys alone. The oldest, Robert, was a student at Knoxville High. He had been happy to be out of class for the summer, though he missed the camaraderie of his fellow cadets on the Junior Reserve Officer Training Corps (JROTC) rifle team, which was one of the best in the entire South. Robert would later join the navy and serve with distinction in World War II, posted to the huge Pearl Harbor Naval Station.

David Leon, known as "Rabbit" for his quickness on the football field, was fifteen when his mother died. Rabbit excelled in basketball and baseball as well as football—quarterbacking the Boyd Junior High Tigers to the city championship against archrivals Park City. He was a gifted musician too. He could pick up virtually any instrument and play it by ear. Later, in the '30s and '40s, he traveled the country playing the clarinet, saxophone, and trumpet for bandleaders like Tommy and Jimmy Dorsey.

His younger brother, Thomas Carlyle, aged thirteen, was artistic and witty. He excelled at graphic arts and acting. As a young man working at the Tennessee Theater, he would pick up the nickname "Shakespeare," or simply "Shake." While cleaning the empty theater between shows, Shake liked to perform monologues, singing and dancing to entertain his coworkers. A decade later Shake would open Edmonds Display Service, where he created hand-painted banners, displays, and marquees for local businesses, like the Tennessee Theater and Miller's Department Store.

As the youngest, Roddie was the most vulnerable, the one son most in need of a maternal figure. I can't imagine what that must have been like for Dad. Only three when his mother died, he must have grieved terribly for her. But as the months and years passed, Roddie learned to understand Jennie's death. Later, he told me that he tried to live a life that would have made his mother proud.

After Jennie died, T.C.'s older sister, Sarah Edmonds Hickman— Aunt Sallie—stepped in to help raise Roddie. Right before he started school, Roddie moved in with Aunt Sallie and her husband, William, a retired carpenter. "Aunt Sallie was heaven-sent," Roddie would say later in life.

Their place on Peddie Street, a dead end off the main thoroughfare of West Blount Avenue, was across the river from the University of Tennessee. If the wind was right, Roddie could catch the smell of fresh bread coming from Kern's Bakery just over the hill on Chapman Highway. Dad used to tell me that he loved waking up to the smell of their breads, rolls, cakes, and cookies. On quiet fall afternoons, he later would often reminisce, one might hear the whistles and shouts from Shields-Watkins Field as the Tennessee Volunteers football team practiced. The Vols, coached by General Robert R. Neyland, would soon be one of the best teams in the nation, and in the mid-1920s, the team loomed larger than life to Roddie.

Roddie loved playing in the big field atop the hill at Fort Dickerson with his neighborhood friends. The fort had been built by the Union army in 1863 across the Tennessee River from Knoxville to prevent Confederates from bombarding Knoxville and driving out the army. Roddie and his buddies played baseball and football or built little shacks on that same hillside.

The homes on Peddie Street were sturdy: made of brick and good lumber. Forward gables extended over the bungalows' wide front porches, which afforded folks a quiet place to take in the neighborhood. Despite their modest dimensions, these front porches were perfect for spending time with family and friends, relaxing in a double swing, or napping in a rocking chair. At the same time, the front yards offered Roddie and the other kids an ideal place to play tag, and the concrete walkways, to ride pedal cars or make a fast getaway on their bikes.

The neighborhood was, in a sense, a broad extended family. Everyone seemed to look out for one another, even my dad, who in 1926 was just a seven-year-old boy. In December of that year, he wrote a letter to Santa Claus that ended up in the evening edition of the Knoxville paper:

I am a little boy 7 years old. I go to school all I can and want you to bring me a car big enough to ride in, a black board, a derrick, lots of nuts of all kinds, and plenty of candy. Please dear Santa bring Boots something nice too. He is a little sick boy who lives with us and he needs an invalid's chair. Don't forget us, we are good little boys. I am a little boy without a mother and Boots is without a father.

Roderick Edmonds.
702 Peddie St.

In the first grade, Roddie brought home a handmade book of the alphabet complete with a backward *S* for "sun" and pages that jumped from *U* to *Z*, leaving out the letters *V, W, X,* and *Y.* He was proud, and so was his family. Roddie would later master his ABCs and begin a lifelong love of word puzzles. He kept that little alphabet book and his dad's pocket-size Bible as the only mementos from his childhood. I would find them years later, tucked away along with his wartime diaries.

Roddie was a dedicated and conscientious student. He also volunteered for the safety patrol and helped serve students and parents. When he was ten, the Great Depression hit Knoxville particularly hard. Following the crash of the stock market in October 1929, Caldwell and Company, the largest bank in Tennessee, collapsed, triggering a financial crisis. It was hard for a skilled wallpaper hanger like T.C. to get work.

Soup kitchens sprang up on Gay Street, and local charities were overwhelmed trying to feed the hungry. Knoxville's growth virtually stopped, as many folks returned to farming or left town. Much of the middle class in Knoxville ceased to exist. The city was forced to pay its employees in scrip and begged creditors to allow it to refinance its debt. Residents abandoned membership in service organizations, disconnected their telephone service, and went hungry. To make matters worse, the three long years of drought that had plagued the South made the dreadful lack of food even more acute. As winter approached in 1932, there were

reports that thousands of people were not only malnourished in
Knoxville but also starving. Roddie and his family experienced severe hunger for the first time. It was a tough lesson, but he learned
it well, appreciating the little he had and how to survive the worst
of circumstances, a practice he continued throughout his life. Dad
knew how to make things last, like his "seegars," a habit of survival he and his generation picked up during the Great Depression. He may not have had the best stuff as a kid, but along the
way he treasured the *right* stuff.

In the presidential campaign of 1930, as Governor Franklin
Delano Roosevelt of New York offered the nation a "New Deal,"
many Americans—especially in Appalachia—hungered for
change. "The hope that had almost ceased to glow, now burns
anew," one Tennessee voter wrote to Roosevelt after his victory.

In May of 1933, President Roosevelt, recognizing the crisis in the
rural South, signed the Tennessee Valley Authority Act. Even by
dire Depression standards, Tennessee Valley was near economic
destitution in 1933. Malaria was rampant, the average income was
under $640 per year, eroded and depleted soil had led to a collapse
in crop yields, and Tennessee's best timber had long ago been cut.

As any schoolkid from Tennessee can tell you, the Tennessee
Valley Authority was designed to tackle the region's economic
woes and, in the process, modernize the region. It developed
phosphate fertilizers, taught farmers ways to improve crop yields,
reforested arid land, and improved the marine habitat for fish
and wildlife. The most dramatic change, however, was from
TVA-generated electricity. Electric lights and modern appliances
made life easier for people, including my family, and helped make
farms more productive. At the same time, the TVA put thousands
of unemployed men to work building dams and other projects.
The electricity also drew dozens of new industries to the region,
providing badly needed jobs. As one Depression-era Tennessee

farmer said, after his religious faith, "the next greatest thing is to have electricity." After I started working for the TVA in college, Dad told me, proudly, that the "TVA was a godsend. It gave us freedom from depression and despair. There's nothing better than freedom."

By the fall of 1933, the fourth year of the Depression, Roddie was an eighth-grader traversing the halls of Boyd Junior High— the same hallways his brothers had walked a decade before. While many older students had to quit school to work or help on the family farm, Roddie was able to stay enrolled. He was one of the lucky ones, because in Roddie's neighborhood, economic conditions would remain austere throughout the '30s. Although 2,500 jobless men and women in Knoxville had gone back to work by April 1933 as part of the National Recovery Administration (NRA), unemployment and hunger remained for many. No one believed the hard times would end.

At Boyd, Roddie excelled at English, science, math, civics, and art. Like his older brother Thomas, he had a keen eye and steady hand for drawing shapes and characters. His favorite subject, however, was history: he discovered that the past was a powerful influence on the present—and the only reliable predictor of future events.

For the most part, Roddie's time in school was typical for that age. He went to school with the kids he grew up with, no more than a mile or so from their homes. He made good marks, generally stayed out of trouble, and participated in clubs and extracurricular activities. Roddie may not have had much in the way of material things, but he lived in a good home, sang in his church choir, and loved baseball—just another seemingly average all-American boy. In that sense, Dad's school-day experiences weren't that different from my own.

But another life-altering moment for Dad occurred at church.

As a young teenager, in the pews of Vestal United Methodist, not far from his home, he became a follower of Jesus Christ. He described it as being "saved"—experiencing "believer's peace," which he recorded in the back of his Bible along with the four verses that helped him understand: Romans 3:23; Romans 6:23; Romans 10:9–10; and Ephesians 2:8–10.

Until then, his sin—his selfishness—had been as natural as his breathing, but he had become unsettled, burdened by the weight of his sins. That's when he realized he couldn't fix his sin, only God could. For him, God was real, the Bible was true, and he was responsible for loving God—*and* others. He bowed and prayed: "Dear Lord Jesus, I know I am a sinner, and I ask for your forgiveness. I believe you died for my sins and rose from the dead. I trust and follow you as my Lord and Savior. Guide my life and help me to do your will. In your name. Amen."

When he stood up, he was a different person. From that day forward, my father's life was transformed.

"I died to myself that day and Christ came alive in me," Roddie would later say. "I was born again, born from above."

IN THE FALL of 1935, Roddie began the tenth grade, his first year at Knoxville High School. Like his older brother Robert, Roddie joined the school's nationally recognized JROTC team. With his fellow cadets, he took advantage of the school's firing range on the top floor, becoming proficient with firearms at a young age.

During Roddie's senior year, more than six thousand Knoxvillians were still unemployed. The Depression was relentless. But Roddie, like most, had learned how to survive.

"Every day Roddie would walk alone to high school with a hard biscuit—many days in his ROTC uniform—crossing the Southern railroad trestle over the Tennessee River," recalled one neighbor, Retha Balltrip.

A long, silent walk—nearly two and a half miles from his home in South Knoxville, past the University of Tennessee football stadium, through downtown, to Knoxville High School on East Fifth Avenue. It was more than an hour walking, each way, with just that single hard biscuit for his lunch.

Roddie's high school years passed quickly. After graduating in June 1938, he got a job working as a stock clerk at Charles E. Hunter and Sons, the company where his father was a wallpaper hanger. He was lucky to find work, because Knoxville didn't offer very much in terms of options, not at the tail end of the Depression. I'm sure Dad wanted to carry his own weight and help out Aunt Sallie. He also needed money to court his high school sweetheart, Marie Solomon, a girl from the neighborhood who was full of life. He had met her at Vestal church, and their fast friendship had quickly blossomed into a romance.

Most likely spurred by his faith and a desire to serve the country he loved, Roddie enlisted in the US Army in March 1941, continuing a long Edmonds tradition of military service. While my

grandfather T.C. never actively served in the military, nearly a year before Roddie's birth, he walked into the local draft board in Knoxville on September 12, 1918, and enlisted in the US Army. He was thirty-seven, with a wife and three boys at home, but he was ready to do his duty for the nation. World War I ended two months later, before he could serve. And Roddie's older brother, my uncle Robert, enlisted in the navy in 1933, at the age of twenty-eight, and would spend more than twenty years serving in the Naval Air Command including during World War II.

Nine months before the attacks on Pearl Harbor, Roddie was sent as a private to Fort Jackson, a sprawling infantry training base in South Carolina, some three hundred miles from the only home he had ever known. During his induction process, he and the other enlistees were handed a neatly typed introduction to military life.

"You are now on your way toward the Induction Station where you will very shortly join the greatest 'team' in the world," it read in part. "Whatever may have been your *personal* reasons for volunteering, the wearing of the uniform will make you a member of the best team of other men like yourself who seek to preserve the American way of life."

Roddie was twenty-one years old.

FIVE

A S THE NEWLY inducted private Roddie Edmonds passed
through the gates of Fort Jackson, he no doubt felt a sense
of pride knowing the base was named after fellow Tennessean
Andrew Jackson.

His good feeling didn't last.

Uniformed soldiers heckled Roddie and the other new recruits
with shouts of "You'll be sorry."

Standing with other enlistees, Roddie raised his right hand and
swore that he would "support and defend the Constitution of the
United States against all enemies, foreign and domestic; that I will
bear true faith and allegiance to the same . . . according to regulations and the Uniform Code of Military Justice. So help me God."

At the time of his enlistment, Roddie was part of the massive
ramp-up in the US Armed Forces. The military was changing at a
lightning pace, modernizing from the horse-reliant force of World
War I and improving training techniques and battle tactics. New
divisions were being created almost weekly as hundreds of thou-

sands of new soldiers registered for service. When Roddie arrived
at Fort Jackson, the base was completing a rapid expansion. It
would eventually encompass 52,000 acres, making it the country's
largest facility for training the infantry.

Following his swearing in, Roddie found himself standing
in line, stark naked, as an assembly line of physicians examined
every part of his body, which according to his admission papers
stood 5 feet 5 inches and weighed a very slight 143 pounds, about
80 pounds lighter than the man I had known.

A nervous recruit near Roddie asked a doctor if he thought he
would pass the physical.

"Soldier," the doctor replied, "you are already in the army."

Dad, now Private Edmonds, was assigned to Headquarters
Company of the 121st Regiment, 30th Infantry Division. He
would be part of the Gray Bonnets, a hard-nosed infantry regiment of the Georgia National Guard commanded by Captain
Charles R. Irwin. The swift and crushing victories of Hitler and
the German Army across Europe threatened the security of the
United States. President Franklin D. Roosevelt and Congress had
ordered the Gray Bonnets of Georgia and all the National Guard
into federal service on August 31, 1940. By the close of September,
the regiment had joined up with the 30th Infantry Division at Fort
Jackson. By March of 1941, the division was well established and
nicknamed the "Old Hickory" division, in honor of President Andrew Jackson. The Germans would later call it "Roosevelt's SS"
for their toughness in the European Theater of Operations during
282 days of intense combat from June 1944 to April 1945.

This company my father joined was one of the country's best.
But he and the other green enlistees were far from battle ready.
Dad and twenty-six other privates, most of whom hailed from
Georgia, began a thirteen-week basic-training program.

Every morning at 0630 hours, Roddie added his voice to the

chorus of young men as they responded to roll call. Every day was like the one before: up early, train for hours, breaking only for meals, then back to bed, exhausted.

For Roddie, the South Carolina heat and humidity were staggering. The pouring rain pelted the men's helmets and drenched their olive-drab uniforms, which made their gear heavier than usual. Regardless of the weather, the men never stopped training—even during afternoon thunderstorms or the occasional tornado, which would bring damaging winds, torrential downpours, and hail the size of golf balls. Their sergeant reminded them that large hail pounding their bodies would be the least of their troubles in battle.

Roddie and his buddies marched constantly—thirty-two miles, nine miles, two-hour speed marches, and two miles "double time"—endurance training that strengthened them for the tough days to come. Roddie competed against the other infantry privates in the hundred-yard speed obstacle course, which had to be completed in full combat gear. The men jumped over a two-foot hurdle, vaulted a four-foot fence, ran a maze of posts and lintels, climbed a seven-foot wall, jumped a six-foot-wide ditch, and crossed a high beam. In another obstacle course, they rope-climbed a twelve-foot wall, then sprinted up a tilted ladder and across a log, before jumping between a framework of planks. All this before swinging over a

water-filled ditch and monkey-barring over another ditch, then crawling through a narrow tunnel and under a wire entanglement.

Simply completing these brutal courses brought Roddie a tremendous sense of accomplishment. Many of the men referred to Fort Jackson as "hell on earth," and to Roddie it was certainly the hottest and most grueling experience of his life. Still, he endured it, pushing through the pain and physical exhaustion. Within a few weeks, he could feel himself getting stronger. The repetitive drills and calisthenics, coupled with the countless push-ups and pull-ups, slowly hardened his soft physique into powerful lean muscle and sinew. He was becoming a proper soldier.

Roddie's training also included intensive study, in courses like Army Organization, Military Discipline, Articles of War, Hygiene, First Aid, Combat Intelligence, Weapons, Mines and Booby Traps, and Close Combat. As part of his education and training, he was exposed to nonlethal agents, like phenacyl chloride—tear gas— and taught to rapidly don his gas mask. He also learned to identify other lethal chemical weapons used during the First World War, like mustard gas.

What Roddie most feared, however, was the "infiltration" course. While machine guns fired over his head, he had to crawl fifty yards under barbed wire. One false move, one nervous raising of the head, and a soldier was dead, which tragically happened on more than one occasion. Every infantryman dreaded it—one guy finished the course with a .30-caliber hole in the mess kit fastened to the back of his pack. Though it was meant to simulate battle, Roddie realized, even as he navigated the barbed wire's sharp points, that as horrific as the course was, nothing could prepare him and the others for the brutal conditions they would face as soldiers in combat.

And yet Dad never stopped training. He quickly distinguished himself on the shooting range as a sharpshooter. His rifle skills, honed during his hours of practice in the JROTC program, earned him the highest classification of rifle expert. He knew his rifle better than anything else. While he could handle all calibers of military firearms, his specialty was the M1 carbine, the lightweight .30-caliber semiautomatic rifle, which he could fire with deadly accuracy.

At the same time, his skill with people and his calm, commanding demeanor won him the respect of both his fellow soldiers and his superiors. Dad was fast gaining a reputation as a natural-born leader.

What Roddie and the other men were training for was never far from their minds. Though the United States was at peace, the war was raging in England. And ten days after Roddie's arrival at Fort Jackson, on March 31, President Franklin D. Roosevelt sped into Fort Jackson at 0900 sharp on an unannounced visit in the "Sunshine Special," his 1939 Lincoln convertible, to inspect the camp and watch the troops parade.

Roosevelt's short visit inspired Roddie, though it wasn't the first time he had been impressed by FDR. On September 2, 1940, six months before Roddie had enlisted, thousands of cheering admirers jammed "flag-draped Gay Street" to see FDR and Mrs. Roosevelt pass through the city on their way to dedicate the Great Smoky Mountains National Park. Boy Scouts kept excited pedestrians off the Gay Street Bridge while a few industrious kids hawked the Stars and Stripes: "Everyone needs a flag! Only five cents! Republicans and Democrats!" Dad had joined the crowd as the presidential motorcade passed.

Only hours before the president arrived in Knoxville, two Royal Navy destroyers had been sunk by Nazi U-boats in the North Sea, and the Luftwaffe had sent more than two hundred bombers and fighters to pummel London and other English cities. During the dedication ceremony at Newfound Gap, a mile high on the Tennessee–North Carolina state line, President Roosevelt offered his dire warning: "The greatest attack that has ever been launched against individual freedom is nearer the Americas than ever before. If we are to survive, we cannot be soft. . . . Squirrel rifles are no longer adequate to defend the nation. We must prepare in a thousand ways."

By the summer of 1941, the US was waking up to a difficult national transition from peaceful neutrality to wartime readiness. The US military needed troops, equipment, and training. In June, Roddie found himself back in his home state, where his Gray Bonnets regiment and 77,000 army troops invaded middle Tennessee for maneuvers. These were the most massive war games in US history, the first time tank, anti-tank, and air forces had participated in large-scale military training. From 1941 to 1944, Tennessee hosted more than 800,000 of the army's finest for similar drills.

Bivouacked at Camp Forrest, near Tullahoma, Roddie's unit took part in various field exercises as they awaited General George S. Patton's Second Armored Division, comprised of approximately 11,000 men and 2,000 vehicles. The "Hell on Wheels" division headed north to Tennessee from Fort Benning, Georgia, in two columns, spanning sixty miles from front to rear. Patton seemed to be on every hilltop and in every valley—leading, yelling, cursing, laughing. Like most everyone else, Roddie viewed Patton as a swashbuckling presence, a strutting commander who regularly brandished his pearl-handled 1911 Colt .45 semiautomatic pistols, a sight to behold.

Equally impressive were Roddie and the other young soldiers. Though food, water, fuel, heat, and hygiene were constant concerns, their poise and good-natured humor helped make these Tennessee Maneuvers memorable. The boys were more than just

US soldiers in training; the citizens of Tennessee embraced them as their own with pride.

The lack of rainfall made dust an enormous problem during moves and marches. When rain did arrive, it brought with it mud. But Roddie and the other boys made the best of it. They also put up with a lack of sleeping quarters. Several thousand soldiers slept wherever they could find space—parks, playgrounds, any place out of the path of jeeps, trucks, and tanks.

In early June, the rigorous training began in earnest. Temperatures hit the upper 90s as Roddie and the other men practiced specific simulations: marching in rapid fashion to gain better ground and surprise opposing troops; staging a counterattack against a larger opposing force; and retreating from a superior number of troops.

One night in June, Roddie and the men of the 30th, 27th, and 5th Divisions moved out in darkness and rain to defend an approximately twenty-two-mile line between Deason and Big Springs. Close-quarters marching at night in secrecy through rain and mud was difficult and taught the men a valuable lesson: weather was always a potential enemy.

Toughened by the long, brutal marches in full gear and the stress of defending positions or advancing an attack, Roddie learned how to hide himself and his equipment from aerial observers, how to move with his unit quickly, and how to conduct harassing rear-guard activity. He also learned that the games were dangerous and deadly. In the first two and a half weeks of combat simulation six soldiers died. The final death toll for the month-long training would be twelve.

Still, the maneuvers helped Roddie grow beyond his high school JROTC and basic-training experiences to more of a well-rounded soldier, one who had endured prolonged bouts of hunger, extreme thirst, constant fatigue, and the relentless stresses of war.

During these exercises, ingenuity—to find provisions, to win the battle, and to summon the will to survive—became the golden rule, the same as in real war.

On June 25, Secretary of War Henry L. Stimson arrived to observe the last part of the games. At 0500 on June 26, Stimson jumped into the seat of a jeep and visited various locations. He was excited to watch two regular army divisions (the Second Armored and Fifth Infantry) take on two former National Guard divisions (the 27th and 30th Infantry Divisions) in the war games.

Like Stimson, nearly everyone in middle Tennessee was awed by the sight of the war games: paratroopers gliding down in farm fields, thousands of infantrymen firing rifles and machine guns at one another—all while tanks rumbled, artillery pieces boomed, and aircraft soared overhead. The rank-and-file soldiers like Roddie took battlefield skills with them through these exercises, while the top leaders made strategic and tactical notes on the best and worst ways to lead men in combat, which helped them determine which commanders were adept at leadership. Roddie stood out among the enlisted men.

Returning to Fort Jackson in July, Roddie made enormous

strides, earning rapid promotions from his commanding officers. By the end of October 1941, after only half a year as a private, Roddie was promoted to private first class. Nine months later, in July 1942, he was bumped up to technician fourth grade—radio operator—and by the end of August 1942, he was promoted to staff sergeant. Finally, on January 19, 1943, Roddie was promoted to master sergeant and communications chief of his regimental company, advancing from raw recruit to master sergeant in only twenty-two months—a virtually unprecedented achievement. A position of that responsibility, the highest-ranking NCO in a brigade, was typically filled by much older career army men. Today it may take fifteen years for someone to make master sergeant, but Roddie achieved the rank by the age of twenty-two.

"Congratulations," Roddie's commanding officer told him during his promotion ceremony. "You're the youngest soldier to make master sergeant."

THE US ARMY was adapting and improving rapidly too. The surprise Japanese attack on Pearl Harbor on December 7, 1941, had killed more than 2,400 Americans and roused a sleeping giant. Across the nation young men—driven by outrage, anger, and love of country—rushed to enlist in the army, navy, and marines. College students delayed their degrees, recent grads left their jobs, and kids in high school too young to enlist lied about their age at recruitment centers. Even veterans of the First World War and old men who had never served in the military were ready to go fight. Within days, the ranks of the military swelled by tens of thousands. Total strangers from every city and village, holler and hillside, bonded as one for a singular purpose: to preserve freedom.

Back in Knoxville, the day after the deadly attack, radio stations WROL and WNOX took out full-page newspaper ads promising twenty-four-hour coverage. Sergeant Alvin C. York, recipient of the Medal of Honor during the First World War—born and raised in nearby Pall Mall, Tennessee—declared, "We're going to give Japan a thorough licking," as he left for Tullahoma on an undisclosed duty for the army.* Many families in East Tennessee clutched pictures and reread last letters from their loved ones killed in the attack. All of America grieved with them.

After news of the attack on Pearl Harbor reached Knoxville, Roddie and his family waited anxiously. Roddie's older brother Robert was serving at the Pearl Harbor Naval Station. Finally, after days of prayer, the family received word that Robert had survived the assault. He would go on to serve with distinction during the rest of the war, fighting in the Pacific aboard the USS *Sirius*, shuttling supplies and Japanese prisoners through the dangerous waters of the Midway Islands.

On December 11, 1941, four days after the United States' declaration of war against the Empire of Japan, Adolf Hitler backed up his Axis partner by declaring war against the United States. Later that day, the United States declared war on Nazi Germany. All told, sixty-one countries would join the global conflagration. The US was now committed to battling fascism on two fronts, taking on the mighty military forces of both Japan and Germany.

In mid-1942, seven months after Pearl Harbor, Prime Minister Winston Churchill secretly traveled by train from Washington, DC—where he had been reviewing war strategies with President Franklin Roosevelt—to Fort Jackson, which was training more than 42,000 soldiers. Churchill's visit wasn't reported in the papers until June 28, after the prime minister had returned

---

* Sergeant York received the Medal of Honor for leading an October 1918 attack on an enemy machine-gun position, which resulted in more than two dozen German casualties and the capture of 132 soldiers. ("Sergeant York, War Hero, Dies; Killed 25 Germans and Captured 132 in Argonne Battle," *New York Times*, September 3, 1964.)

to England. "Winston Churchill saw a spectacular display of America's expanding might Wednesday at the army's largest infantry training post, where crack paratroops plummeted from the sky by the hundreds and live ammunition from big field guns whistled directly over his head and burst near enough for him to feel the jar and concussion," reported *The State*, the newspaper of Columbia, South Carolina. Chomping on one of his Romeo y Julieta cigars, Churchill "inspected Fort Jackson's activities minutely, even prying into soldiers' packs, working the breech block of a 75 millimeter gun, and getting covered with choking, yellow dust kicked up by thousands of feet and hundreds of armed vehicles. He saw some of the plain, essential drudgery of life in an army camp."

Roddie and the other soldiers gained a new focus and a sense of their future allies from the charismatic visitor from England, while Churchill, weary from war, found renewed strength and optimism from the might of the US infantry. Watching the thousands of recruits undergoing training, Churchill said, "They're just like money in the bank."

On September 30, 1942, President Roosevelt made another surprise visit to the fort. He'd been on a secret tour of military installations by train, crisscrossing the country from Michigan to California to Texas to South Carolina and points in between. He visited thirty facilities in eleven states in only fourteen days, never venturing far from the tracks. The president traveled more than 8,500 miles by train and motorcade through twenty-four states under the cover of government secrecy to get a firsthand account of military production and citizen morale while the US was in the throes of war in Europe and the Pacific.

Along the way, Roosevelt saw impressive displays at the Chrysler tank plant in Detroit; at the first inland navy training center in Lake Pend Oreille, Idaho; in Vancouver, Washington, at the Alcoa

aluminum plant; at Camp Pendleton, the United States Marine
Corps base in San Diego; at Higgins Industries' boatyard in New
Orleans; and at Camp Shelby in Mississippi. "And yesterday," the
president told reporters, "we turned up at Fort Jackson, just outside of Columbia, South Carolina, and reviewed another division,
which was in a different stage of training from any that we had
seen before." The commander in chief had been impressed by the
preparedness of both citizens and soldiers—none more evident
than the troops of Fort Jackson.

SIX

AT NOON ON March 15, 1943, a gleaming limousine pulled up in front of Fort Jackson's Theater 2, carrying the governor of South Carolina, Olin Johnston, who took his place on a stage next to Major General William Simpson, commanding general of the XII Corps.

Onstage sat Brigadier General Alan Walter Jones along with the senior staff of a new division about to be created. In the body of the theater were cadres—including Roddie and 1,800 men drawn from the 80th Infantry Division—who had arrived over the past few days. The average age was twenty-one, including all officers, and an experienced group of men with Roddie from the 80th.

The 106th Infantry Division, which had been constituted on paper in May 1942, was then formally activated, with 16,000 personnel from nearly every state in the union except the states along the Pacific Coast. Allied High Command had recognized the need for more land forces in preparation for the eventual invasion of

Europe, and the 106th, an exceptional group of soldiers who had scored high on intelligence tests, would be the last infantry division created during the Second World War. The division adopted the motto "To make history is our aim."

Afterward, as the units left the theater, the commanding officer of troops turned to his adjutant and recited some words of poet Richard Hovey: "I do not know beneath what sky, nor on what seas shall be thy fate: I only know it shall be high, I only know it shall be great."

Three days later Brigadier General Alan Walter Jones was promoted to major general and appointed commander of the 106th Infantry Division, known as the Golden Lions. The divisional insignia was a yellow lion's face on a bright blue background encircled by white and red borders. The blue represented the infantry; the red, the supporting artillery; the lion's face was symbolic of strength and courage.

Jones was a stocky fifty-year-old with "jet-black hair, heavy eyebrows, and a pencil moustache." He was regular army but hadn't attended West Point military academy. He'd graduated from the University of Washington, where he'd studied chemical engineering on an ROTC scholarship. In World War I, he was commissioned as a second lieutenant with the 43rd Infantry Division but never saw combat. As training began in earnest, Jones would say, "Everything seemed to be going our way and the world looked bright and cheery."

Roddie, for one, was thrilled to be in a new division and in charge of fresh men, ready to take the fight to Hitler. On March 29, 1943, as the sun began splashing fiery hues over Columbia, a bugle blared, and training began for the 106th Infantry Division. Inspections, the presentation of colors, and breakfast were over by 0700 hours, when training of the 16,000 fresh warriors began with rigor. Roddie was tasked with transforming these raw recruits, and he did his job well, but he also understood the awesome duties in front of him.

Roddie had changed rapidly since his enlistment. To his fellow infantrymen, he seemed mature beyond his years. Now a newlywed, Roddie was looking forward to a post-army life back in Knoxville as a good husband and provider, and to building a family. A few months earlier, in late November 1942, while temporarily stationed at Camp Forrest with the 319th Infantry Regiment, Roddie had slipped away to Knoxville on a weekend pass, and that Saturday night, in an intimate ceremony of family and friends, he had married his girlfriend, Marie Solomon, a petite local beauty from his neighborhood. Three years his junior, Marie was stunning as she walked down the aisle of their church, Vestal United Methodist—the place where their friendship had begun and blossomed.

After graduating from Knoxville High in 1941, Marie had gotten a position as a clerk at the Vestal Lumber and Manufacturing Company not far from her home on Scottish Pike. The Vestal community, founded by the Vestal family when the lumber company opened, was once an incorporated city known as South Knoxville. The company had situated itself around Maryville Pike and Ogle Avenue, where other businesses then located, forming a downtown area. The lumber company employed hundreds of people and was, at one time, one of the largest in the South. It was a good job for Marie, and she would continue working while Roddie was away serving the country.

The wedding had passed quickly, and before he knew it, Roddie was back at camp leading his men. By 1943, as a master sergeant and communications chief in Headquarters Company of the 422nd Regiment, one of three regiments of the new 106th, his duties included installing and operating wire, radio, and air-to-ground communications, and directing transmission and receipt of calls and messages, which were vital to his entire division. His most important task was instructing and training his men in the techniques of field communication. On the battlefield, of course, communication was critical. Roddie also was responsible for the physical transformation and education of the men of the new division. He had them running a hundred-yard obstacle course, wearing full combat gear—jumping over hurdles, fences, and a six-foot-wide ditch, and running a maze made up of posts and lintels—competing for the fastest time.

When he wasn't running his men ragged, Roddie also made sure to teach them to love and respect their rifles. Most of the infantrymen carried the M1 Garand .30-06 caliber, the workhorse rifle of the Second World War, unlike Roddie, who still preferred the M1 carbine he'd mastered during his own infantry training. By putting his men through drill after drill, Roddie had them so familiar with the Garand that they could fieldstrip and reassemble their rifles while blindfolded.

Roddie had trained many young recruits, and one of his best was Lester J. Tannenbaum.

LESTER WAS BORN on August 8, 1923, in the South Bronx, where his family lived in a modest apartment. Lester's father, Louis, had emigrated from Austria in 1911 and ran a small tailoring business on the Lower East Side. During Lester's youth, the Bronx was a

thriving hub for immigrant Jews, having more residents of Jewish descent than any other borough. By 1930, nearly 50 percent of the borough was Jewish, about 364,000 concentrated in the South Bronx and in the apartment houses along the Grand Concourse, the four-and-a-half-mile boulevard sometimes called the Champs-Élysées of the Bronx.

Neighborhoods like Lester's—his parents originally had lived in the Highbridge section—teemed with Yiddish-speaking merchants, kosher butchers, ritual bathhouses, and synagogues. The Tannenbaum family kept Orthodox traditions. Lester followed in the footsteps of his older brother, Paul, becoming a bar mitzvah at age thirteen. Their mother, Frieda, kept a kosher home; her third child, Corinne, arrived in 1926.

For the Tannenbaums, as for the Edmonds and millions of other Americans, the Great Depression was the defining challenge of their time. "I was six years old in 1929 when the Depression started, and I was seventeen years old in 1940 when it ended," Lester told me in one of our many conversations. "Those eleven years were the cruelest years for the American people. I happened to be fortunate because my father was a businessman. He wasn't an employee during the Depression. Of course, his business was not very good, but he had a contract making uniforms for the policemen and firemen of New York City. He had a shop [that] had maybe nine or ten workers, mostly working on sewing machines. And his business was hurt because the police and firemen weren't spending as much money. But he always had some money coming in, and it kept us going. . . . My father and mother managed to keep the family together and happy."

Lester's family never wanted for food. And his father's shop had enough spare cloth from the uniform contracts to be able to manufacture clothing for the family.

"I wore his suits. He made suits for the whole family—my sister as well. So we were some of the lucky few. We were clothed. We had food. And we had a place to stay. We kept moving around in the Bronx because every time you moved you got a one-month concession. We moved often. It wasn't a big deal. I lived in three different apartment houses—on Walton Avenue—on the same block."

Lester's father had arrived in the United States at age fourteen, and like many immigrant boys, he went to work right away. He didn't have a formal education, but he taught himself to be fluent in English and spoke without much of a discernible accent. Part of that education was reading the *New York Times* every day. "On Sunday we'd gather—that one day the whole family would gather in the living room and the *Times* would be spread out. Everyone would be reading a section. My father loved this country. He had come from persecution—he'd escaped persecution. That's why he had us all reading the *New York Times*—a very strong voice on democracy and what is good about our nation."

I can't imagine a more different childhood experience from Dad's. While Lester and his family had been dressed in custom-made clothes and were reading the *Times*, Dad likely had been making good with what he had, mending his clothes and reading the *Knoxville News Sentinel*.

Lester recalled that his older brother briefly "flirted" with a youth group connected to the Communist Party when Paul was still in high school. "One Sunday he told my father about a meeting he was going to that night. For the first time I saw my father become truly angry, and as my brother left, my father threw his shoe at him. He said, 'You're crazy! You live in a wonderful country. You're gonna support a Communist Party?'"

Lester was an exceptionally bright student, and at the Bronx's Public School 33 he accelerated several grades so he was ready to enter high school by age twelve. His parents had set the academic bar high for all their children. "At that time, the only elite high school in New York was Townsend Harris," Lester recalled, a three-year high school for which "you had to pass a citywide test. I was one of the hundred who were accepted. It was a very traditional, classics-based school." In addition to learning Latin, students were required to study a second, modern language. "I took German because I thought I was going to be a doctor." At the age of fifteen, Lester started classes at City College—then known as the "Jewish Harvard."

Lester's brother, five years his senior, was already in the service—ultimately commissioned as a major with the army's Transportation Corps—and Lester was anxious to complete his coursework as quickly as possible to go for a commission in the US Army Air Forces. His dream was to be a fighter pilot, in the cockpit of a Mustang or a P-38 Lightning, terrorizing Messerschmitt fighters over the skies of occupied Europe.

He had started out as a pre-med student but disliked medical studies and quickly changed to pre-law. "And when I turned eighteen—in August of 1941, four months before Pearl Harbor—I registered for the service. I knew that I was going into the army, but I was still in my program at City College, and you were deferred until graduation. I was a junior at the time. After Pearl Harbor, if you were in college, you were exempt from the draft until you finished your degree. I was supposed to graduate in June of 1943, but I accelerated my program, took extra credits, so I could graduate in June of 1942, a year ahead."

After receiving his bachelor's degree, he was immediately inducted into the US Army at Fort Dix in New Jersey. "I felt great, couldn't wait," he said. "If anything, I was afraid I wouldn't be able to get into the war. I was nineteen years old. When you're nineteen, [you think] nothing can happen to you."

At Fort Dix, despite his preference to be a pilot, the brass pegged him for the infantry, assigning him to Fort Jackson. "The reason I went there was they were putting together a brand-new outfit—the 422nd Infantry Regiment."

Lester arrived at Fort Jackson in early March 1943 and found himself immediately assigned to Headquarters Company of the regiment, where he met Roddie, the top NCO in the command-and-control regiment, and the leader and instructor of the GIs.

"The regiment had five thousand enlisted men, three battalions," Lester recalled. "Each battalion had three rifle companies. And then there were special companies like Cannon Company, Anti-Tank Company, Heavy Weapons Company, Supply Company. The regimental headquarters was responsible for directing all these units. The colonel would take his assignments from the general at division headquarters, and it was up to the colonel to implement those assignments. He would send out his commands to the various battalions, and my job during basic training was to be at the command post."

Most of the top commissioned officers—unlike Major General Jones—were West Point graduates. And the junior officers—first lieutenants, second lieutenants, captains—were often referred to as "ninety-day wonders," because they had completed Officer Candidate School, or OCS, in just three months. The "top noncommissioned officers," Lester remembers, "were all regular army. They had been in the army, peacetime army, and been trained. They were there before Pearl Harbor and they were training us. As an enlisted man you have a lot more respect for the peacetime army guys than the ones that come in—whether drafted or volunteered—just during wartime."

Lester spent the first thirteen weeks of his basic training under Roddie's leadership. "Roddie was my commander and later my close friend. He combined the unusual virtues of courage and sensitivity. He knew the army's code of conduct very well, which required him to take care of the men under his command to lead the infantry."

As the ranking NCO, Roddie was the right-hand man of the 422nd regimental commander, Colonel George L. Descheneaux Jr. According to Lester, Roddie's word "carried a lot of weight."

In the South Carolina heat, Roddie turned green boys like Lester, from wildly diverse backgrounds, into tough, skilled infantrymen. He took enormous pride in the job, and his men respected him. Throughout the regiment, Roddie was known for his calm but firm demeanor. He never threw his rank around. He was tough but fair—his sole aim was to prep the young recruits under his command for combat conditions.

What struck Lester most about the curly-headed master sergeant from Knoxville was that Roddie always taught by example. If they were on the firing range, he showed the men how to take up a proper stance and aim. He also showed them how to carefully pack their backpacks so they could survive in any situation. And he led them on marches: marches for endurance, speed marches, double-time marches. No matter the distance, no matter how sweltering the temperature, Roddie accompanied his men on every march. He trained them for the final infiltration course too. Just as Roddie had done, Lester and the other new men had to crawl under barbed wire as machine guns fired live rounds over their heads. It was a daunting test for many of the men, but it was necessary for them to develop the courage to face live ammunition and to face their fears before graduating from basic training.

First and foremost, though, Roddie taught his boys the importance of following orders, to respect the chain of command, be-

cause just as he had been taught during his training, failing to do so would likely get them killed.

Another of these green young recruits was a charismatic young Italian American from New York named Frank—or Frankie, as everyone called him—Cerenzia, who would become Roddie's closest confidant. Like Roddie, Frankie was a married man. He had wedded his beloved Lucy days before leaving for basic training. Maybe that's what helped my father and Frankie form such a fast friendship. They both understood the pressures of being enlisted, married, and separated from their brides. Marriage in the army was difficult at best, and it seems that Roddie and Frankie supported each other. Or maybe it was their belief in God. While Frankie was a devout Catholic, he and my father, a Methodist, both expressed a sincere faith and shared a common view of life. They saw people, problems, and purpose in the same light, I suppose.

Frankie was born in Brooklyn on April 23, 1924, the son of Gabriele Cerenzia, a blacksmith who had emigrated from Campari,

Italy, and Anna Devona, a girl from New York City. Living on DeKalb Avenue in the Bedford-Stuyvesant neighborhood of Brooklyn, Frankie attended two years of high school at Erasmus Hall before enlisting in the US Army on February 8, 1943.

"I gave Frankie his basic training and he was one of my best soldiers," Roddie wrote in his diary. "We became very close and had a lot in common. He's an alright guy, and my best friend—truthful and sincere. I always admire a man who is truthful."

Roddie, Frankie, and Lester became extremely close at Fort

Jackson. Though Roddie had rank over the other two men and was clearly in command while on duty, off duty he was relaxed and lighthearted, a good friend to all the men, especially Lester and Frankie. During their free time, Dad and his "boys" listened to Glenn Miller, the Dorsey Brothers, and Benny Goodman. I'm sure Dad couldn't help but brag about how his older brother "Rabbit" played with a lot of the big bands. They also enjoyed Count Basie, Duke Ellington, Ella Fitzgerald, and Frank Sinatra—music lifted their spirits.

During the hot, hazy summer of 1943, training was intense. Fierce rains fell, pinging helmets and stinging bodies. Roddie, Lester, Frankie, and the others endured it together. There was something almost spiritual in their unity. They found energy—a second wind—to push them through, and they experienced comfort even as steam rose to meet them from the hot South Carolina sand.

Their tight friendship, however, was soon split up. Roddie recommended Lester for the infantry school at Fort Benning, where he would spend the next three months training to become a noncommissioned officer. "Follow me"—the Infantry School's motto—was precisely what Roddie had taught all the new recruits to do under his command: leading not by barking out orders or stern repetition but by personal example. After Lester's training, he applied for the US Army Air Forces and was sent to Fort Bragg, North Carolina, the site of USAAF headquarters. There he was put through a battery of tests and was ultimately ordered to Stuttgart Army Airfield in Arkansas, where he would qualify to be a fighter pilot on the Lockheed P-38 Lightning, a complex twin-engine, turbo-charged aircraft armed with a 20 mm cannon and four .50-caliber machine guns.

While Lester learned to be a "flyboy," back at Fort Jackson, Roddie, Frankie, and the rest of the 106th continued their intensive training, aiming to make history.

During his secret visit to Fort Jackson the previous year, Winston Churchill had foreseen the division's potential; he had told Roddie and the ranks of infantrymen engaged in a mass calisthenics exercise, "I know you are all waiting and longing for the day, which is coming, when all this work and preparation will be turned into a mighty effort of war to make sure that right and justice will prevail in the world."

SEVEN

I N 1943, IN OHIO, another young Jewish American, named Sydney Friedman, was selected by the US Army as potential officer material.

Sydney—known as "Skip" to his family—was born in Cleveland on April 24, 1924, to Louis and Pearl Friedman, immigrants who'd come from Eastern Europe in 1922. Louis had fought for the Austro-Hungarian Empire in the First World War, was captured by the Russians, and was sent to a remote POW camp in Siberia. After his release in 1917, Louis made the long overland trek back to his home in Poland, and along the way he met and married a beautiful blonde named Pearl Rosenbaum before they set out together for America. While most of Louis's family had already emigrated, Pearl would be the only one of her family to leave Europe—the rest stayed behind and would perish in the Holocaust.

The couple sailed for New York from Antwerp in 1922, and after a brief stay in the coal country of Pennsylvania, where Louis had relatives, they ultimately settled in Cleveland. Louis opened a small business, Atlas Cleaners, and the family established themselves in the community.

Skip was very bright and athletic, like his older brother, Morrie. At John Adams High School, Skip played defensive end on the varsity football team. Morrie was already a star football player there by the late 1930s, and the pair often made news in Cleveland.

Skip aspired to a career in journalism, and while still in high school, he worked as a stringer for the *Cleveland Press* newspaper. Morrie left Cleveland on a full scholarship to play football at the University of Alabama, and in 1942, during his senior year of high school, Skip was offered a scholarship to play football at DePauw University in Indiana. But with the US now in World War II, and the draft age lowered to eighteen, Skip realized he and Morrie were going to be called to serve, so he turned down the scholarship offer—one of the "great disappointments" of his life—and enrolled at Case Western Reserve University. Eight months after he turned eighteen, he entered the service along with Morrie.

"We took basic infantry training at Camp Wheeler, Georgia," Skip recalled. "It was very tough and rigorous training. Then out of the blue my brother and I both ended up in something called ASTP. I'd never heard of it before. They told us 'ASTP stands for Army Specialized Training Program.' What the army had in mind for us I had no idea, but there we were, in the ASTP."

Though Skip had not heard the name, the Army Specialized Training Program had already, by late 1942, begun transforming the character of the US military. With the US now fully at war, the army had launched the "largest educational program in the nation's history."

Immediately after Pearl Harbor, many universities and colleges nationwide faced an uncertain future. Most of their students who were twenty years old had enlisted or were on the verge of being drafted. To make matters worse, the Selective Training and Service Act of 1940 had lowered the draft age to eighteen. As colleges worried about dwindling enrollment, the army realized that, in the event of a protracted war, the military might run short of well-educated officers and technical specialists.

The ASTP was the army's solution. Loosely modeled on the modest Student Army Training Corps of World War I, the program was approved by Henry Stimson, Roosevelt's secretary of war, in September of 1942 and fully implemented by December. The idea behind ASTP was to cull the nation's best and brightest from high schools and colleges around the country. Some candidates were as young as sixteen, but eligibility was based solely on intellect and previous educational success: candidates had to be high school graduates and score a minimum of 110 on the Army General Classification Test, a Stanford-Binet-type intelligence scale, or 10 points higher than what was required for Officer Candidate School.

Colonel Herman Beukema, a professor of history and government at West Point, ran the program, insisting on a dictum of "soldiers first, students second." Though ASTPers weren't yet in active combat, their service in the program was no cakewalk. They wore regulation uniforms and were subject to strict military discipline and inspections; participated in all normal formations, including reveille; marched to classes and meals; and had mandatory lights out at 10:30 p.m. Plus, they were prohibited from participating in any intercollegiate sports. Each week included twenty-four hours of classroom and lab work, twenty-four hours of required study, five hours of military instruction, and six hours of physical training.

There had never been anything approaching the scale and ambition of the ASTP. The US Army was, in effect, attempting to fast-track the training and education of hundreds of thousands of soldier-scholars. All advanced degrees were accelerated: four-year coursework was crammed into twenty to twenty-four months. Aside from the students in medical, dental, electronics, and veterinary programs, most ASTPers were enrolled in engineering, foreign language, and basic psychology courses. The soldiers were stationed as units on many college campuses around the country.

The vast scope of the program—and the implicit whiff of elitism—made it immediately controversial. While some colleges went out of their way to welcome the soldiers, a few barely tolerated them as a revenue source. General Lesley J. McNair, an old-school West Point man and World War I vet, was the commander of US Army Ground Forces by 1942. He felt the ASTP took young men with leadership potential away from combat positions, where they were most needed. Others claimed the army brass had gone along with the program only to keep the universities from protesting to Congress and the Roosevelt administration about the lowering of the draft age. One congressman derided the ASTP as a "gambit designed to keep the sons of the rich and powerful out of combat," allowing them to "loaf and play" on campuses "while the sons of honest working men perished on the field of battle."

As baseless as those allegations of favoritism were, ASTPers were nonetheless seen as an elite and erudite cadre within the US Army. On the rare occasion when they did intermingle with regular infantrymen, the ASTPers were mocked as geeks, bookworms, and the nickname that ultimately stuck: "Quiz Kids." Even in their olive-drab uniforms the Quiz Kids were easy to spot. All ASTP students wore a unique shoulder patch depicting a blue "Lamp of Knowledge," which was sometimes derisively referred to by grizzled old soldiers as a "piss pot."

Today, the ASTP is regarded as perhaps the boldest and most far-reaching educational experiment in US history. All told, Colonel Herman Beukema, who later asserted before a congressional investigating committee that the ASTP was more rigorous than either West Point or the US Naval Academy, would be responsible for sending 200,000 soldiers to 227 land-grant colleges at a cost of $127 million. Those soldiers would eventually include US senators Bob Dole and Frank Church, US secretary of state Henry Kissinger, New York City mayor Ed Koch, and numerous prominent writers, entertainers, and artists, such as Gore Vidal, Kurt Vonnegut, and Mel Brooks.

The ASTP would leave a profound footprint on US history and culture of the twentieth century. It has been called a groundbreaking social experiment in merit-based advancement, one that helped to more fully "democratize American society by selecting its trainees based on their inherent ability rather than on their family's socio-economic status."

SKIP FRIEDMAN FOUND HIMSELF in the Deep South after his time at Case Western Reserve University. Like his older brother, Morrie, Skip was enrolled at the University of Alabama in Tuscaloosa. "We were involved in a West Point kind of engineering program," he told me at his home in Shaker Heights, Ohio, an idyllic community outside Cleveland. Regina and I had driven up from Tennessee to meet Skip after Lester had connected us. He proved to be a wonderful storyteller, and I quickly understood why he won over so many people. He was funny and wise and open, and his stories helped me piece together more of my father's story.

As Skip told me about his wartime experiences, he was surrounded by family: his daughters, Amy and Rachel, his son, Peter, a few grandkids, and their extended families. For some of them, it was their first time hearing about Skip's experiences during World War II.

In spring 1942, as he was departing for Alabama from Cleveland Union Terminal, Skip met a group of coeds who'd come to "say goodbye to the boys." Among them were two young Jewish women, Penny and Ruth, the Klausner twins.

"Penny and I formed a quick connection," Skip said. "As we boarded the train, Penny looked up at me with her nice eyes and said, 'We'd like to know what's going on in the army during these wartime days. Why don't you write us and we'll write back?' So

I quickly took out my pad and wrote her address down. And we started writing letters. That fateful meeting, our correspondence, and two dates before I shipped overseas sustained me through the terrors and sorrows of war."

IN 1944, WHILE SKIP was in the ASTP, the 106th was slogging hundreds of miles to and from the war games in the Tennessee hills, preparing for a long trek to its final posting in the Midwest before shipping out to Europe. On January 20, under the command of Colonel Walter C. Phillips, Roddie and the 422nd Regiment led the trek, in a truck convoy, from Fort Jackson, South Carolina, to Camp Atterbury, Indiana, an arduous six-hundred-mile trip, which took the regiment nearly three months to complete. In the succeeding days, the two other regiments of the 106th followed, the 423rd Regiment, under Colonel Charles C. Cavender, and the 424th Regiment, under Colonel J. L. Gibney.

The men of the 106th made themselves at home in Camp Atterbury, and in the months ahead, Roddie would continue to train his troops for the upcoming land battle in Western Europe. Now, though, his training took on a new urgency, as the ranks were being depleted by a steady need for reinforcements in other divisions. Through the course of the war, no other infantry division would lose as many men before entering combat as the Golden Lions. Especially after the heavy infantry losses the Allied forces incurred during the amphibious landing and invasion of Italy in 1943, the 106th Division at Camp Atterbury was regularly called upon to provide freshly trained "replacements."

Between September 1943 and its entry into combat, the 106th lost 12,442 men—80 percent of its total. As the official history of the Army Ground Forces notes, "the withdrawals were often made in driblets, aggravating the disruption of training. For example, there were fourteen separate withdrawals, 25 to 2,225 men at a time."

Congress, in 1944, set the army's size at 7.7 million and the navy's at more than 2 million for the impending amphibious invasions of Europe and Japan. (By 1945, the US Armed Forces would number an astonishing 15 million men and women in uniform.) As the Allied High Command began top-secret planning for the June D-Day invasion, the US Army decided it needed millions of frontline fighters rather than the ASTP's cadre of college-trained specialists and officers. The old guard found the idea of the army educating soldiers ludicrous. Generals like Lesley J. McNair, a longstanding critic of the ASTP, argued that they needed *riflemen*, grunts on the front line. McNair scoffed, "And we're sending men to *college?*"

On February 18, 1944, the Army Specialized Training Program was halted at all campuses nationwide. Most of the ousted ASTPers went into infantry, airborne, and armored divisions still in stateside training but scheduled for shipment overseas before the end of 1944. Some thirty-five divisions got an average of about 1,500 men each.

If the ASTPers expected any special treatment, they quickly found out otherwise. Despite their intellect and credentials, few gained the chance to attend Officer Candidate School, and practically none got noncom ratings until they reached combat zones, where heavy casualties created vacancies for them to fill. Also, the ASTPers often received a harsh welcome from the regular infantrymen, who considered them a bunch of "smart-ass" college kids needing to be taught a lesson about the rough reality in the "real army." As one company commander put it: "What kind of soldiers deal out bridge hands during their ten-minute training breaks?"

Roddie saw things differently. As he met the young ASTPers, like Private Skip Friedman, he immediately recognized that these boys had quick wits and expertise in areas like communications and engineering, which could be an advantage in a combat situation. He felt many of the ASTPers were "officer material" and his division was lucky to have them.

"Your father was a pleasant surprise," Skip told me. "Some of the noncoms were shouters, but Roddie was different, determined, and steady as a rock. He was a terrific guy. Quiet, and reserved, just a very steady guy. You could count on him; he was totally dependable. He led by example."

Throughout 1944 Roddie helped rebuild the 106th Division with replacements like Skip from the ASTP and Lester, who arrived back under Roddie's command from the Air Corps. Lester had been deep in training as an air cadet on the P-38 fighter in Arkansas, a few months away from achieving his dream of becoming a Mustang pilot who would shoot down Messerschmitt fighters in the skies over Europe. But the Allied invasion of Europe would change everything.

"We lost so many men during the D-Day invasion—and the demand for replacements from the infantry was so high—that the army decided to take all men who had not completed training in the Air Corps and reassign them to combat units," Lester recalled. "They sent me back to Camp Atterbury, and I was reunited with the 106th Division—with Headquarters Company of the 422nd Regiment. And at that time, the division was already on maneuvers in Indiana—the final stage of training before preparing to go overseas. I got there in time to participate in the last stages of maneuvers. I was no longer an air cadet, but I was promoted to sergeant in the infantry."

At Camp Atterbury, Lester was happy to be reunited with Roddie and Frankie as the division was quickly brought up to full strength. And Roddie and Frankie were happy to have Lester back. Lester had been one of Roddie's best soldiers, and he and Frankie followed Roddie's lead in welcoming the ASTPers, whom they found to be bright and more than capable soldiers.

When the men weren't training, they'd drive up to Indianapolis, where they enjoyed hearty meals at local restaurants, watched John Wayne movies at the Circle Theatre, or visited Riverside Amusement Park. Lester told me that he and some of his friends regularly beat a path to the USO, where they danced with local girls. Once, to everyone's surprise, Vaughn Monroe staged a personal performance with his band, singing his hits "Moon Over Miami" and "Mule Train." Dad wrote in his journal, with characteristic succinctness, that he and Frankie "liked going out together and having good times."

Anything was better than training.

By the summer of 1944, 60 percent of the 106th Infantry Division's enlisted strength—more than seven thousand men—had been detached and reassigned to other units for immediate overseas service, and all the division's new men needed to be trained rapidly, a nearly impossible job no matter how high the caliber of the regimental officers and NCOs. Now a seasoned drill instructor, Roddie tried to prep the men in his command for the inevitable transatlantic voyage to the European Theater of Operations and the gruesome realities of combat. Though still only twenty-four, he nevertheless understood that no matter how well he trained his men, they would likely have no idea what to expect during the brutality of land warfare in Europe.

But Roddie was coping with an unexpected personal crisis too. Early that summer he had received divorce papers from his wife, Marie. "When my wife divorced me, I almost went crazy," he wrote. "Everyone told me it was best, but I didn't care what everyone thought. I only knew it wasn't right with God."

On July 30, Roddie was granted ten days to travel back to Knoxville in hopes of patching things up, but the divorce was made final, on August 5, 1944. He cut his furlough short by five days and returned to Fort Jackson. "I had lost what I had wanted all my life, a home, a wife, and happiness," he wrote. "I was a pretty weary guy when I got back to camp after my wife had thrown the books at me. Frankie seemed to know what was going on in my mind and he tried his best to make me forget."

EIGHT

I ARRIVED IN WASHINGTON, DC, on October 29, 2013, on a fact-finding mission; without ever planning to do so, I was becoming a history detective, spending hours at the National Archives in College Park, Maryland, and at the Library of Congress in Washington researching more about the 106th Division and its heroic role in the Battle of the Bulge.

I found page upon page that I couldn't look away from—it was like some real-life thriller—jigsaw-puzzle pieces from German stalag records and postwar testimony from newly liberated POWs. I'd had no idea, when I first read Dad's diary, how cruel the Nazi regime had treated US prisoners of war. I guess I'd seen too many sanitized Hollywood movies or naively assumed that the standards of the Geneva Convention had protected guys like my dad—and the tens of thousands of other GIs in Nazi camps—from brutal mistreatment.

It startled me to realize how mistaken I was. Slowly, inexorably, I began to see those stark images in my dad's diary as part of a broader canvas. As a POW at Ziegenhain, my dad wrote at one point, "I'm just a little guy," but I could see now that he'd been swept up in an epic and defining moment in the twentieth century. He'd arrived in Europe during the ferocious final months of the war, during the unprecedented genocide against the Jewish people, an era of unspeakable crimes against humanity and a dark time when following one's moral compass, when choosing right over wrong, could prove lethal. While I had learned about World War II and the Holocaust in school, the ruthless nature of Hitler and his Nazi regime was, piece by piece, becoming more visceral, more frightening, more *real*.

What had happened in Germany in the 1940s was no longer some remote slice of history; it was *personal*. What had happened a lifetime ago was my dad's story—and now it was becoming my story too.

I stared at documents and microfilm, took hundreds of pictures on my cell phone, read until my eyes blurred. I knew that studying testimony would fill in many blanks, but the only way I would fully come to understand more about my dad's wartime service was to track down the handful of men, still living, who had vivid memories of serving in the 106th Infantry Division with Dad, of fighting in the Ardennes Forest, of spending time with him in the Nazi POW camps.

What I was most excited about during that trip to Washington, DC, was that I'd lined up a meeting with a former POW named Paul Stern, who lived with his wife, Corinne, in Reston, Virginia, about a thirty-minute drive east of the nation's capital. I knew that Paul—like his good friend Lester—was another vital link, an eyewitness who'd known my father well.

Before I left Washington, I asked him if I could videotape our entire conversation. Paul agreed, and Paul and Corinne's daughter, Joanne, would be there as well. (Joanne would later tell me that she'd never heard some of the details her father told me during the interview.)

In the weeks before my visit, Paul had been very sick, even hospitalized, but when I arrived, I found an eighty-nine-year-old man who, though frail, was alert, animated, and clearheaded. He was dressed in a black sweater and a white oxford shirt and wore tinted wire-rimmed glasses. As I scrambled to get my video recorder going, Paul rushed to tell me of his experiences in the war, his arrival in France, only weeks after the D-Day invasion.

"Our job was to help the wounded on the battlefield, to save these soldiers," Paul recalled. "Shells dropping all over. People getting killed all around you. It was just a different life, but you had to adjust to that life. And not to say that we got *used* to it—it was scary—but you had to adjust to it. Within twenty-four hours, Chris, you were changed. You were a different person."

Like Lester, Paul was born in a heavily Jewish enclave of the South Bronx, on January 27, 1924. His father and mother, Max and Jennie, had emigrated from Austria. The close-knit family— including Paul's older brothers, Jacob ("Jack") and Aaron, and his twin sister, Mildred—lived on Lowell Street in the Bronx, and Max had steady work as a tailor, even during the worst days of the Great Depression.

Paul loved playing stickball in the street in front of his house, not a dangerous pastime in the days when few cars raced around the Bronx. He was extremely bright, skipping a grade in elementary school. Paul was also outgoing, charismatic, and a born optimist who was always ready to find humor in even the darkest situations.

Along with his mother, Paul listened to Metropolitan Opera broadcasts on WQXR radio every Saturday afternoon, likely sparking his love of classical music. He had a strong tenor voice, and at James Monroe High School he sang with the German glee club and was thrilled when he got the opportunity to perform with them at Carnegie Hall. He would never lose his passion for classical music, especially *The Magic Flute*, and he was proud to tell everyone that he shared a birthday with Mozart.

After graduating high school a year early, at age seventeen and near the top of his class, he earned a scholarship to Wesleyan University but chose to stay in New York and attend City College. At eighteen, he enlisted in the service, and on July 14, 1943, he was sent first to Fort Dix, in New Jersey, before moving on to Camp Grant, in Illinois, where he began training as a dental technician. Right away, however, he was fast-tracked to be a combat medic.

"Guys with the highest IQs they put in [to be] medics," Paul recalled. "On the battlefield, in an emergency, you had to be able to think fast. There was a big difference going from being a dental technician to a combat medic. We were so young. We came into action, eighteen- and nineteen-year-olds, and we were about to be thrown on[to] the battlefield."

Paul arrived in France on July 29, 1944, and was assigned to the 28th Infantry Division. Considered the oldest division-size unit in the US Armed Forces, the 28th was known as the "Keystone" because it had originally been a Pennsylvania National Guard division, with some of the units tracing their lineage back to the Revolutionary War and Benjamin Franklin's battalion (the Pennsylvania Associators). Paul and the other teenage replacements felt like interlopers in an old-boys' club, and as combat medics, they were immediately thrust into the chaos of the killing fields of France.

If a medic was lucky, he picked up some tricks of survival from infantrymen with a few weeks' more experience in combat. Paul learned the only way to survive under Nazi bombardment: "We would jump into a shell hole because we knew another shell wouldn't come to the same hole. You were safe in that hole. We learned that almost right away. It was a strange thing, jumping into a fresh shell hole to be safe."

US Army medics were under constant enemy fire yet didn't carry any arms themselves. And they were, in a very tangible sense, marked for death. "We had a red armband. If the Germans wanted to shoot us, they could just aim for the red armbands." In his early days in France, Paul became good friends with another combat medic from Chicago. "We had a wounded soldier on the litter. I picked up the back end, and my buddy was in the front. And as we rose, a sniper shot my friend through the head. Like that—just snuffed him out. I had to write to his parents."

From Normandy, the 28th Infantry Division began a lightning-fast eastward thrust toward Paris in July of 1944, known as Operation Cobra. The joint British, Canadian, and American effort was meant to secure the whole of Brittany as well as the deepwater port at Cherbourg. Along with cutting off the Brittany peninsula from German forces, the success of Operation Cobra would allow US forces to escape the large hedgerows of Normandy that restricted their use of armor. The 28th snaked through the bocage, finding the roads littered with abandoned tanks and bloated, stinking corpses of men and horses. One of the priorities of the mission was to take Caen—Allied commanders considered control of that city vitally important to making a march on Paris. As the operation drove German resistance south, US troops liberated town after town on their way to the French capital.

On August 29, 1944, Paul and the rest of the 28th Division participated in one of the most fabled moments of the entire war: the liberation of Paris. They marched in a stunning olive-drab stream down the Champs-Élysées to the Arc de Triomphe. "We were chosen to be the division to parade through Paris," Paul recalled with pride. "They later put it on a postage stamp. As we paraded under the Arc de Triomphe, you could see that Eisenhower was right on top with de Gaulle and all the generals. It was quite a scene. That evening we got to this park, and we slept on the ground as usual."

Few of the million-plus French people cheering in the streets knew that Paul and the rest of the 28th were already on their way back into ferocious combat. "It was one of the most remarkable attack orders ever issued," General Omar Bradley later said. "I don't think many people realized the men were marching from parade into battle." But within a day, the 28th Division was back in the fury of the front lines, engaging the German positions and driving inexorably toward the Belgian border.

"We chased the Germans the next day," Paul remembered. "And we kept chasing them. At various times we'd engage the enemy, and we'd have a quick fight. One time we were attacked and there were six German half-tracks defending this French village. As we approached, they just turned around and left. Another thing about each village: as we came to it, all the girls that had collaborated with the Germans were brought into the village center and they cut their hair." Paul said he saw women like that in every French town the 28th Division liberated. The collaborator women were "disgraced" and "embarrassed."

"We chased the Germans across France, and we just kept on chasing them." Finally, by early September, he said, "we had some peace. I asked my commander—we were resting—if I could go see my brother Jack. He was in southern France. I got to Strasbourg in one day; the second day I got to his division and I found him. Imagine! I found one soldier in the *entire* division. No one would believe it. I was so driven. I found Jack. I have pictures of the two of us. We were both together in uniform. We were together for a few days. Then I went back to the 28th."

To Paul and his brother Jack, and to most of the other GIs, given the rapid liberation of France and the string of Allied victories, it certainly seemed like the war was almost won.

NINE

As the summer of 1944 turned to fall, the Allied powers became complacent. Paul Stern with the 28th Division saw firsthand that the Nazi panzers and infantry were retreating rapidly in France—driven back into the heart of the Reich—and were being routed by the Soviets on the Eastern Front. After the liberation of Paris, General Eisenhower set up the Supreme Headquarters Allied Expeditionary Force (SHAEF) at the luxurious Trianon Palace Hotel in Versailles just outside Paris.

Eisenhower's chief of staff, Lieutenant General Walter Bedell Smith, told the press in late summer that "militarily, the war is over." The sentiment was widespread. While meeting with President Roosevelt in Quebec, on September 10, Winston Churchill announced:

Victory is *everywhere*. I would not be surprised, now that the American Third Army is standing on the border of Germany, if the enemy surrenders in weeks.

On September 11, in fact, US forces breached the famed fortifications of the Siegfried Line and were, for the first time, occupying German soil.

"Victory fever," it was called.

Of course, it was a false euphoria. The rapid military success
led to major problems: both General George Patton and Britain's
Field Marshal Bernard Law Montgomery, racing each other to be
the first commander to take Berlin, had outrun their supply lines.
Trucks supplying Patton's tanks, for example, had to travel 360
miles back to the Normandy beaches to refuel. Also, a deepwater port like Antwerp was crucial, so Montgomery ordered the
First Canadian Army to take Antwerp. It was one of the fiercest
conflicts of the liberation campaign—the Battle of the Scheldt—as
Germany's 15th Army dug in, fighting fiercely to keep the Allies
from capturing any deepwater port near the Reich.

Still, despite the supply-line problems and intense opposition,
by autumn 1944 Germany seemed on the verge of collapse; Field
Marshal Montgomery had made a bet with Eisenhower that his
21st Army Group would be marching through the ruined streets
of Berlin by Christmas.

It was easy, then, for many of the young soldiers still in training camps or on maneuvers in the US to think that the European
conflict might end before they were deployed. Irwin "Sonny" Fox,
from Flatbush, Brooklyn, was one. Sonny remembered that when
he heard about the D-Day invasion, while still at camp in Louisi-
ana, he said, "Damn, the war will be over before I can catch up
to it."

Sonny was a quick-witted, six-foot-three, rail-thin, first-
generation Jewish American kid from Newkirk Avenue in the
heart of Flatbush, Brooklyn. He loved stickball and streetcar trips
to Ebbets Field to see the Dodgers with his dad. After graduating from Erasmus Hall High School, Sonny started studying the-

ater at NYU and was already hosting his own college radio show when he got the call-up to the US Army.

"I got the draft notice shortly after I turned eighteen, and I reported in for the draft exam," Sonny recalled. "And they found me totally acceptable—I was breathing."

In fact, he was more than just acceptable. Sonny scored high on the aptitude test, and the ASTP selected him for fast-track basic training at Fort Benning, Georgia. Never having been outside New York, Sonny was shocked by the color line in the Deep South, seeing black and white infantrymen unable even to ride the same buses on R&R trips to Columbus, Georgia. "The Great War to save democracy," he wrote, "was being fought by a totally segregated US Army."

Sonny was sent to England aboard the commandeered *Queen Elizabeth* luxury liner. The grand ship was "double loaded" with twenty thousand infantrymen. Eighteen infantrymen squeezed into cabins built for two. After the ship arrived in the Firth of Clyde, Scotland, Sonny continued on to the town of Crewe, in England, a crucial transportation hub in the heart of the Midlands. From England, he made the crossing into liberated France and then into Belgium. That fall, he wrote a letter back to his parents in Brooklyn:

OCT. 23, 1944

Dearest Folks,

I'm sitting right now on Belgian soil (and my own posterior) under a shelter of pine limbs, next to a warm fire, in the coals of which are roasting some potatoes. I've just finished a good hot meal, augmented substantially by one of the occasional cow or deer which "accidentally" strayed in the line of fire of one of the men. Of course, if someone should say I could go home, there would be a puff of dust and a flash of speed and Brooklyn here I'd come!

Sonny said he "finally caught up with the war" at a "repple depple"—as the GIs called replacement depots—along the German-Belgian border, where brand-new platoons were being cobbled together, "new meat to be chewed up at the front," as Sonny described it.

"The policy of the military in World War II was to feed in replacements as the war wore on," Sonny continued. "Occasionally, a whole new division would show up—one that trained together. The men would know their officers and each other and have some cohesion. When we were lined up that night to enter the ranks of Company E, 110th Regiment, 28th Infantry Division, not one of the men in this particular gaggle had ever met the other."

Due to his eleven months in the army, Sonny was considered "experienced"—and given the position of squad leader, in charge of eleven men. He was nineteen.

Like Paul Stern, Sonny now wore a uniform with a red keystone, the official emblem of the state of Pennsylvania, on his shoulder. The Germans, out of grudging respect, referred to the 28th Division as the "Bloody Bucket" because of that crimson keystone insignia and the division's fierce fighting during the Normandy campaign.

Before pushing on into Germany, the 28th and other US forces were given orders to sweep away any resistance from the area of the Hürtgen Forest. Nothing could have prepared Sonny, Paul, or any of the men for Hürtgen. Even the seasoned older German officers later called the fighting in Hürtgen "worse than fighting in the First World War."

"The Hürtgen Forest sat along the German-Belgian border, about fifty square miles of dense woods, with tall fir trees that blocked the sun," Sonny recalled. "It was a dark and eerie place where the thick lower branches of the fir trees were only two feet off the ground. The forest floor never got any sunlight."

The attack by the 28th Division started on November 2, 1944. The German defenders were ready and well positioned; the US troops were immediately pinned down by mortar and artillery fire and harassed by local counterattacks. Just one mile was gained after two days of savage fighting, and the harsh winter weather grounded any tactical air support.

Sonny had arrived at the front in the middle of the night, then climbed a hill and heard other soldiers shouting about "those goddamned 88s!" The Germans' 88 mm artillery was, the men found, as accurate as rifle fire from a great distance.

"As we were walking up to the first positions under fire," he wrote later, "I felt a combination of fear and elation. Fear of what might happen—elation that I was okay. That first walk up there was the first time I'd ever been under fire. I found out that I could make it. When my squad moved into its position in the Hürtgen, we weren't even told where the enemy was. My first job was to take the frozen body of a dead American soldier out of the slit trench I was to occupy. I didn't stop to think about what I was doing. I just did it. After all, I had to settle the squad in. Being responsible for eleven other guys was very important to me. It was raining and cold, a damp, penetrating cold that went through your bones. Nighttime was frequently below freezing, and damp and fog was everywhere. I had one change of socks, which never dried so my feet wouldn't stay warm."

Ernest Hemingway, too, was in the thick of the battle, as a war correspondent, against all regulations, toting a Thompson sub-machine gun and carrying two canteens: one filled with schnapps and the other with cognac. Well into middle age at that point, he mockingly called himself "Old Ernie Hemorrhoid" and knocked out dispatches for a paper in Paris, though he was embedded with the infantrymen so he could gather material for a World War II novel he was planning to write.

Staff Sergeant Jerome David (J. D.) Salinger was just a half-mile away with the 12th Regiment, 4th Infantry Division. He later said that when he entered the heavily fortified Hürtgen, he "crossed into a nightmare world." Somehow Salinger managed to find time to write short stories—which he sent to the *New Yorker*—throughout the battle, "furiously writing," he said, "whenever I could find an unoccupied foxhole."*

Perhaps the most terrifying aspect of that "nightmare world" was the shelling tactic the Nazi artillerymen had perfected. "The Germans had timed the fuses on their shells to explode at treetop level," Sonny recalled, "thus dispersing the deadly shrapnel over a wide area. A substantial number of our casualties were from this shellfire."

"In the Hürtgen Forest," Paul Stern said, "the way the shells would go off, hitting the treetop, spreading the shell, it could kill you when you were half a mile away. The shell and fractured tree parts would spread out. It was *hell*. We thought we were all dead men. Every day we woke up in the forest we didn't know if we would make it to the next day."

The battle of the Hürtgen ended in a major German defensive victory.

---

* The experiences J. D. Salinger endured in Hürtgen are essential to under-standing his later life and work—most clearly seen in his semiautobiographi-cal narrator Sergeant X in the short story "For Esmé—with Love and Squalor" (*New Yorker*, April 8, 1950).

For the GIs, it was relentless and horrific warfare, lasting five full weeks, all resulting in a losing effort. During the battle, 33,000 Americans were killed or wounded—an astonishing 25 percent casualty rate. Sonny and Paul's 28th Division suffered 6,184 combat casualties, plus 728 cases of trench foot and 620 cases of "battle fatigue"—or PTSD, in today's terminology.

Hürtgen is viewed by historians as among the worst Allied debacles of the war, and many survivors never spoke of the battle again.

MONTHS BEFORE THE 28th Division was mired in the Hürtgen Forest, the Nazi state's full racist ideology finally was exposed when, on July 23, 1944, Soviet soldiers captured the first of the Nazi death camps—known as Majdanek—located in eastern Poland. Initially, the exact nature of Majdanek was a mystery to the Red Army troops; seeing the camp's high walls, large metal gate, brick buildings, smokestacks, and barracks, the Soviets had assumed it was a kind of factory. Indeed, it was—a factory of death. When the Red Army soldiers opened the doors to a gas chamber, they found a crematorium that was still warm from burning corpses.

It took many months before a disbelieving world came to grips with the news. First, Soviet war correspondent Konstanin Simonov was flown into Lublin to write a series of articles about Majdanek in the Red Army newspaper *Krasnaya Zvezda*, then the *New York Times* sent its veteran war correspondent W. H. "Bill" Lawrence to Lublin so he could see the evidence for himself. Lawrence's stunning report ran on the front page on August 30, 1944, under a banner headline:

Nazi Mass Killing Laid Bare in Camp
  Victims Put at 1,500,000 in Huge Death Factory of
  Gas Chambers and Crematories

For millions of Americans, it was the first news account of a crime so vast, grotesque, and industrialized that it still had no name.* For the first time the world heard journalists describe Nazi gas chambers disguised as showers, canisters of lethal cyanide-based Zyklon B, and enormous piles of shoes—more than 800,000 pairs overflowing the camp warehouse. After the discovery of troves of passports and identification papers, Simonov revealed that many of the murdered had been small children and the elderly, and that an enormous number of Jews had been "brought to the camp to be exterminated from literally every country in Europe, from Poland to Holland."

Bill Lawrence's lead in the *Times* was stark and unequivocal: "I have just seen the most terrible place on the face of the earth—the German concentration camp at Ma[j]danek, which was a veritable River Rouge for the production of death, in which it is estimated by Soviet and Polish authorities that as many as 1,500,000 persons from nearly every country in Europe were killed in the last three years." Holocaust scholars now believe the number of victims at Majdanek was far less than those initial estimates; still, at least several hundred thousand civilians and Soviet POWs had been executed in the death camp; on a single day in November 1943 the Nazis, who kept meticulous records, slaughtered 18,400 people in Majdanek.

BACK AT CAMP ATTERBURY, Roddie was leading the men of the 106th Division through intensive drills, including long-distance lake swimming, preparing them to move overseas that autumn.

---

* For decades the Nazi genocide had no proper designation in English. Many historians opted for "the Final Solution," a direct translation of the euphemistic phrase the Nazis themselves coined. In Hebrew, the murder of millions of Jews became known as "the Shoah"—or "the Catastrophe"—used as early as spring 1942 by the Jerusalem-based historian Ben-Zion Dinur to describe tragedy befalling European Jewry. Only in the 1960s did English-language scholars begin using the term "Holocaust."

Roddie and the others undoubtedly heard reports of the Nazi atrocities against Soviet prisoners of war and civilians in eastern Poland, which must have steeled their resolve. This was no longer simply warfare but a clash of ideologies, a fight to save the world from murderous tyranny. And the Jewish GIs at Camp Atterbury, young men like Lester and Skip, realized that by wearing dog tags that identified their religion, according to US Army policy— stamped *H* for Hebrew—they were at immeasurable risk if they should ever fall into Nazi hands.

Before the division was shipped out, Major General Alan Walter Jones addressed his troops: "Be proud of your assignment, of the fact that you have been selected for a combat division, that on your shoulders rests the responsibility for the victory we have to win. Never forget that your individual part is of first importance to the success of the division."

On October 11, 1944, the 106th Infantry traveled by train to Boston and then, on October 20, on to New York City, where volunteers from the Salvation Army handed out kit bags filled with combs, toothbrushes, toothpaste, and "other much-appreciated goodies" on the Hudson River piers. Aboard the RMS *Aquitania*, Roddie and the 422nd and 423rd Regiments left New York Harbor on October 21, 1944, bound for the European Theater.

"Our ship was large and too fast for the stodgy Merchantmen escorted by the Navy," recalled John Morse, a sergeant from the 423rd Regiment. "She had guns mounted fore and aft and a British crew. Our days, about a week, were spent in long, serpentine chow lines, coiling, morning to night, about the deck of the ship. Loaded to capacity the ship drove into the Atlantic mists as we loitered in line for a breakfast of [porridge], made with sea water—we vowed—to dinner of mutton stew and bread. Some of the Brit crew regaled us about how our officers, above deck, were eating like kings. They generously offered to sell us steak sandwiches—for a dollar a pop, American. As we edged along the chow lines we griped, for the most part, and moved our card games steadily forward."

As the ship neared the continent, the British crew began firing their deck guns. Someone had spotted a German U-boat, but the *Aquitania* made it safely through the Firth of Clyde and docked at Gourock, Scotland, on October 29, 1944.

Like Sonny, the men of the 422nd traveled south, by train and truck, into the English Midlands. There they rendezvoused with the rest of the divisions at the Grange at Guiting Power, outside the small city of Cheltenham. They would be stationed at "Guiting Grange" for a month, gathering equipment, drilling, and training. They heard educational talks from US combat veterans who'd already seen action against the Nazis in France. And they enjoyed socializing during their time off from training.

"Cheltenham's a small community," Lester recalled, "and there were no men—just women and young girls. We were invited to parties—USO had parties every weekend. That's where I learned how to do the hokey pokey."

The men also could get a pass and go to the movies. "One time I had a date with a girl from Cheltenham," Lester recalled, "and we went to the movies. As soon as the film starts, somebody comes on stage and says, 'There's a bomb in the theater.' I expected to evacuate the theater, but the guy says, 'Everyone look under their seats and see if there's a bomb.' And they went around, inspecting the building. It turned out to be a false alarm, and they resumed the movie." Lester was aghast. "My date, she expected that would happen. They were so used to unexploded bombs. Those were the conditions the British people were living under."

By November 29, 1944, the 106th Infantry Division had been assigned to VIII Corps, First Army, 12th Army Group. They would reach French soil, it seemed, by early December. The officers of the 106th were told that the men were being given an "easy" post-ing, being sent to a sector so uneventful that GIs had nicknamed it the Ghost Front.

According to the official *Army Morning Report* of November 30, 1944: "Departed by rail 0130 for Southampton. Weather cold and clear. Morale good."

PART III

*I can remember thinking that it would be simpler,
and more effective, to shoot [the replacements] in the area
where they detrucked, than to have to try to bring them back
from where they would be killed and bury them.*

—ERNEST HEMINGWAY

TEN

———

THE 106TH DIVISION'S orders were to take up position on the front lines where, by the first week of December 1944, all significant fighting had ground to a halt. The giddy optimism of the early autumn had vanished; despite the seeming hopelessness of their position, the Nazis were putting up much more of a fight than anyone imagined possible after the Allied success at Normandy and the liberation of Paris in August.

Operation Market Garden—launched in September 1944, the strategic brainchild of British Field Marshal Bernard Montgomery—had turned out to be an unmitigated failure. It had been a daring Allied offensive fought primarily in the Netherlands and the largest airborne operation in history up till that time, with the objective of capturing nine crucial bridges that might have provided a clear route into Germany. While the Allies had succeeded in liberating the Dutch cities of Eindhoven and Nijmegen, at the Battle of Arnhem they were unable to capture the last bridge over the Rhine. In the nine short days of Market Garden, some 17,000 Allied troops were killed, wounded, or missing in action.

The failure of the operation coupled with the horrific carnage in Hürtgen Forest had brought a grim change of mood. The lightning successes of commanders like Patton and Montgomery meant the Allied supply lines were dangerously overstretched, and the lack of access to deepwater ports presented Allied High Command with enormous problems in keeping frontline troops supplied with food, fuel, arms, ammunition, and winter clothing. Using the Normandy beaches to land provisions was inadequate to meet operational needs.

The only deepwater port the Allies had taken was Cherbourg, but the retreating Germans had mined the harbor and destroyed the docks. The Belgian port of Antwerp then became the key objective, and the fighting for the city was ferocious. In the first days of September, the Allies had captured Antwerp's port intact, but it was not fully operational for landing supplies until late November.

The previously complacent Allies faced a stark reality at the end of November 1944: there was no chance of reaching Berlin and ending the war in Europe by Christmas. Indeed, it was clearly going to be a long fight to bring down the Thousand-Year Reich. The war on the Western Front had become a slow battle of attrition against an increasingly desperate, fanatical, and battle-hardened enemy.

After months of being deflated by the constant news of losses on the Eastern Front, the Allied advance after D-Day, and the growing sense of defeatism in the Reich, Hitler claimed to have developed a plan for a devastating counterstrike. Back on September 16, 1944, Hitler confided in General Alfred Jodl, his principal executive officer, that he had made a momentous decision. "I shall go over to the offensive! We'll break out of the Ardennes, with the objective Antwerp!"

A replication of the May 1940 Blitzkrieg? Hitler's plan certainly wasn't lacking in audacity. The German Army would drive through the dense forest of the Ardennes and cross the River Meuse, then sweep north to retake Brussels as well as the deepwater port of Antwerp. Cut off from their US allies, the Second British and First Canadian Armies under Montgomery's command would be enveloped and destroyed. Political tensions were already rising between the British and Americans, and Hitler reasoned that continued pressure would collapse the Western alliance against the Axis, freeing Germany to deal with the Red Army on its Eastern Front. Several of Hitler's generals protested, but the führer would not be deterred from his last great strategic roll of the dice. When Hitler unveiled his plan to Field Marshal Karl Gerd von Rundstedt, his commander in chief in the west, the old Prussian was staggered. "It was obvious to me that the available forces were far too small—in fact, no soldier really believed that the aim of reaching Antwerp was really practicable," he later said. "But I knew that it was useless to protest to Hitler about . . . anything."

Even the fanatically loyal Oberst-Gruppenführer Sepp Dietrich, commander of the Sixth Panzer Army, sarcastically complained: "All Hitler wants me to do is to cross a river, capture Brussels and then go on and take Antwerp. And all this in the worst time of the year—when the snow is waist-deep . . . !"

The officers of the German High Command tried to explain the enormous manpower advantage held by the United States Army. But Hitler didn't think the US Army could *fight*. Given the inherent racist ideology at the heart of Nazism, it was impossible for Hitler to understand that the *diversity* of the United States, of divisions like the Golden Lions, was, in fact, its greatest strength.

IN DECEMBER, THE ALLIES stalled at the German border. Thousands of GIs and British soldiers were positioned at the edge of the Reich. Before them lay the Siegfried Line—the Reich's formidable last defense before the Rhine River—some 430 miles of concrete pillboxes, barbed wire, tank traps, foxholes, and interlocking fire zones. All along that front, also known to Germans as the West-wall, things were static. And brutal winter weather was arriving.

The Germans dug in along the Siegfried Line and waited for the big Allied attack—most likely in the Ruhr Valley, the Reich's

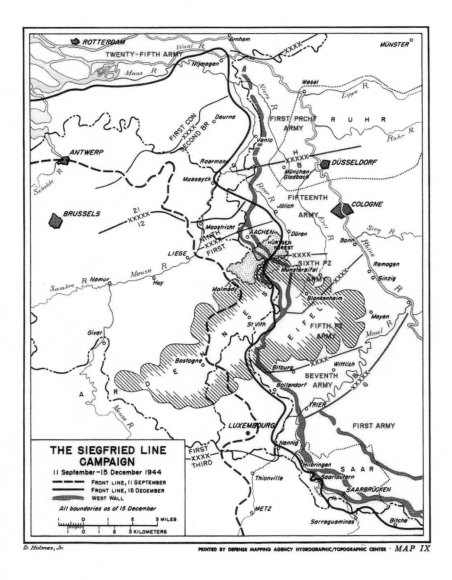

THE SIEGFRIED LINE
CAMPAIGN
11 September–15 December 1944
——— FRONT LINE, 11 SEPTEMBER
——— FRONT LINE, 15 DECEMBER
▰▰▰ WEST WALL

All boundaries as of 15 December

D. Holmes, Jr.

PRINTED BY DEFENSE MAPPING AGENCY HYDROGRAPHIC/TOPOGRAPHIC CENTER · MAP IX

industrial heartland. Their forces on the Eastern Front braced for the massive winter offensive still to come from the Soviets—expected as soon as the muddy earth of Prussia froze solid enough for the Red Army's tanks.

There were sporadic artillery barrages, machine-gun bursts, and an occasional sighting of a German combat patrol. "Social calls," the GIs jokingly termed them. But that was the extent of the action, even though parts of the US Army were occupying German soil.

It seemed to many observers like the kind of entrenched stale-mate characteristic of the First World War.

ON DECEMBER 2, 1944, Roddie, Lester, Frankie, and the other men of the 106th Division made landfall in France, disembarking at Le Havre on the Normandy coast. Each man climbed down rope netting onto landing barges, which pitched and heaved in the cold North Atlantic surf. Each man staggered under the weight of his equipment and weapons: "Full field packs, helmet, rifle, loaded cartridge belt and an additional two bandol[i]ers of ammunition," in the words of Sergeant John Morse of the 423rd. "This 'sheet' weighed as much as we did. Our barge didn't get far enough ashore so we all jumped into the surf and staggered ashore."

That night, Lester wrote a letter to his older brother, Paul, who was serving with the US Army Transportation Corps back home in New York City. Lester wanted Paul and his family to know he had arrived safely in France and that all was well. His words were optimistic and encouraging—just as he intended. Though he would be heading to the front soon, he assured them he would be stationed in a "quiet sector." No cause for alarm. And besides, Lester wrote, the French seemed like "fine people and fun to know." Since most were saying the war would be won by Christmas, he looked forward to enjoying his mother's onion rolls and cream cheese after taking care of some business with Hitler.

Shivering and soaked, the men of the 106th began a long trek toward the front. They traveled in uncovered trucks, through freezing rain and snow, across France and Belgium. Between December 8 and 10, all regiments of the division were scheduled to reach the forests of the Ardennes plateau near the ancient town of Saint Vith in Belgium, where the Second Infantry Division had been guarding that sector of the front.

It was a harsh introduction to life in the field; the men of the 422nd, 423rd, and 424th regiments were freezing and wearing drenched uniforms and boots by the time they arrived at Saint Vith.

On the pitch-dark night of December 9, after another all-day motor march in their open trucks, through a blinding snowstorm, Roddie's 422nd Regiment gathered at an assembly point outside Saint Vith. The heavy snowfall and the restricted use of lights made it very difficult for the men to follow their guides. Some men were separated from their units and became completely lost in the dense woods.

Roddie and his men found themselves positioned almost exactly at the center of the Allied front: in the heart of the Ardennes—the sprawling hilly woodlands that rolled through parts of France, Belgium, Germany, and Luxembourg. It was an "old-growth pine-forest, in places still primeval." Only a scattering of small towns could be found in a predominantly "vacant landscape of trackless woods, rolling hills, steep ravines and ridges."

The 106th's orders were to take up positions on a long, thin front just outside Saint Vith, along the Schnee Eifel. According to an after-action report written by Major William P. Moon, the Schnee Eifel was "a hogback ridge, characterized by high plateaus, deeply enclosed valleys, and a restricted road net. This area [was] heavily wooded with steep ravines and ridges running to the east and west. The road net consisted of narrow dirt roads in poor condition running generally north and south along the ridge. Due to the snow and ice at this time some of the roads were almost impassable."

Although common military wisdom labeled the Ardennes "impenetrable," armies had in fact traversed it for centuries—from ancient Roman times to Charlemagne's, and indeed Hitler's blitzkrieg attacks of 1940. The locals in the area were mostly of German descent with a sprinkling of French-speaking and Flemish Belgians; as the Allied soldiers would soon learn, many towns had divided loyalties, some with German-speaking pro-Nazi citizens, some with active members of the Resistance.

When Roddie and the men of the 106th arrived at the Schnee Eifel, intermittent flurries were continuing to limit visibility while a thick mist hung over the treetops. The Schnee Eifel may have seemed like a serene Christmas-card landscape to them, with its fog-shrouded hills and its pine trees covered in fresh snow. To Roddie, it eerily resembled the hills and hollers of East Tennessee.

WHEN THE 106th GIs piled out of their muddy trucks near Saint Vith, they were met by dozens of bearded, dirty infantrymen— ragged and stinking men who hadn't washed in weeks and were desperate for a hot shower. The 106th had been ordered to swap places with the exhausted Second Infantry Division, man for man, across a twenty-eight-mile sector. To the rookie troops, the men of the Second Division looked almost savage, grinning knowingly at the newly arrived kids. Some of the veteran infantrymen offered sarcastic and profane best wishes as they piled into trucks to be taken away from the front lines.

"Lucky guys!" other battle-wise veterans shouted. "You're coming into a rest camp."

"It's been so quiet up here," one regimental commander of the Second told the regimental commander of the 106th, "your men will learn the easy way."

Roddie himself was told by his CO that the first few weeks on the Schnee Eifel would "be a piece of cake." By late 1944, the area was even said to have become a "paradise for weary troops."

Everything was indeed quiet in their sector of the Ghost Front. Almost *frighteningly* quiet.

This was precisely the reasoning behind General Omar Bradley's decision to call up the inexperienced 106th Division to the Ardennes—they needed an undemanding post where green troops could gradually adjust to wartime life. "New arrivals needed a chance to ease into their new assignments, make friends, feel a part of something, and learn how to survive in combat," wrote historian John C. McManus. The 106th had the distinction of being the last US infantry division to be mobilized in World War II. It was also the youngest—the first division into combat with substantial numbers of eighteen-year-old draftees. Two-thirds of its troops were single men under the age of twenty-three. The Ardennes, Allied High Command reasoned, was the ideal place for their acclimatization.

In the cold gray dawn of December 10, Roddie and the men of the 422nd Regiment dug themselves out of their snowy bivouac area. Roddie ate a "skimpy" hot breakfast, then companies were formed and began preparations for the move to the front lines.

According to a regimental after-action report: "The disposition of troops for the 106th Division was to be elements of the 14th Cavalry Group and the 422nd Infantry Regiment on the north, the 423rd Infantry in the center, and the 424th Infantry Regiment, less one battalion, on the south. One battalion of the 424th was to be in division reserve."

Transportation in the forest was clearly going to be a problem.

The roads through the Ardennes were "in poor condition, covered with snow and mud." The men took up positions in foxholes, mostly prepared by the Second Division, and largely filled with mud and water. All terrain was potentially treacherous: "There were mine fields and barbed wire entanglements out to the front. Trip flares and antipersonnel mines were scattered along the front in the gaps between occupied positions."

Skip Friedman recalled that they were right on top of the Nazi fortifications. "We were among the first troops into Germany proper . . . on the German side of the border. We were in the middle of the Siegfried Line, in their own tank traps, bunkers, all kinds of defensive fortifications."

Life at this fortified wintry front had myriad hazards, beyond being killed by enemy gunfire or land mines. The winter of 1944/45 would prove to be one of the coldest and wettest ever recorded in Europe. Trench foot was one of the awful frontline realities that the 106th had yet to experience—veterans of winter combat had learned to stave it off only through firsthand experience and the advice of other seasoned troops. Unlike frostbite, trench foot did not require subfreezing conditions, merely damp and unsanitary ones. The painful disease—technically, an "immersion syndrome" in which erythema or cyanosis as a result of poor blood supply results in necrosis of the toes and flesh of the foot—came from simply not changing into dry socks regularly.

If a soldier's feet remained immersed for long periods in cold water, they'd turn blue, begin to blacken, and become gangrenous. Untreated, severe trench foot would often lead to amputation. Being a green division, with no experience in winter combat, few of the men of the 106th understood the necessity of clean, dry footwear, and the Second Division while rotating out had taken the heaters on which they had been drying their own wet socks. Before the men had even settled in on the Schnee Eifel, they experienced casualties due to the extreme weather. Dozens of troops had to be evacuated to the rear, suffering from trench foot.

"Both the enemy and the weather could kill you," one private observed, "and the two of them together was a pretty deadly combination."

As promised, the first few days in the Schnee Eifel were routine for Roddie and the rest of the division. Then came ominous rumblings. The men kept hearing the distinctive noises of tanks, trucks, and other armored vehicles, but their warnings were downplayed as an overreaction of a green division that had yet to fire a shot on a battlefield. But even the much more experienced Fourth Division, positioned directly to their south, "assumed that the engine noises came from one Volksgrenadier division being replaced by another."

The loud noises from the German side of the line increased each day, including the whistle of steam locomotives carrying troops and materiel across the Prüm Valley. At Corps headquarters, no one appeared concerned—even after Luftwaffe recon planes were heard flying over their positions. All reports sent up the intelligence chain were chalked up to "nerves"—jumpiness by the new arrivals.

As the communications chief for Headquarters Company, Roddie was responsible for radioing in reports about the heavy-equipment noises and suspected troop movements, but his commanding officers continued to downplay them. When the 106th's commander, Major General Jones, told his superiors at VIII Corps headquarters he himself was hearing armor, he got a blunt reply: "Don't be so jumpy, General." The men were even told that the Germans might be playing recorded sounds of tanks over loudspeakers to spook them.

Of course, the sounds were all too real. For this last manic gamble, under cover of heavy, cold fog, hidden in the thick forest, the

Germans had amassed seven panzer and thirteen infantry divisions for the opening attack.

Two days before the brutal battle began, Paul Tannenbaum responded to Lester's letter:

THURSDAY—14 DECEMBER 1944

Dear Les,

Received your letter written from France on the 4th of December. Your letters are cheerful and full of good spirits and one can't feel badly long when you get so much of your energy into the written word. The mention of "onion rolls and cream cheese" has stirred Mother to action. In most of the future packages that she will send out you will probably find close facsimile of either or both. She is kind of worried, though, over the fact that you haven't received any of the other packages. I guess they will catch up with you in time—although Mother is inclined to believe they fell into the hands of the enemy and because of these extra provisions the war is being prolonged.

Of course, the news that you are in France . . . is not conducive to happy feelings. But sitting down and thinking it all out logically—a hard task—there's not much one can do—except fight all the harder to make this damn war business over and done with. There's a great deal of living for you to catch up with and it irks me something awful to see you wasting years . . .

But from your feelings toward the French people already expressed in your letter, I don't imagine you will be lacking fun in your free moments—which I know will come far and in between from here on in.

I haven't had much time to write a long letter from the office as I did in the long, long past—but as soon as the work slackens a bit you'll get those daily typewritten sheets again. Meanwhile, I'll use those postcard V-mail forms—just to keep my hand in. Anything from you is a treat and if you have some stuff from England that you are cluttering about—you can send it to me at the base. Also, anything that you feel you are carrying around in excess—send it back to me. And should you need something don't hesitate to ask for it.

Keep well, Les.

Always,

Paul

Paul Tannenbaum's letter would never reach Lester in the Ardennes.

ELEVEN

HITLER HAD WANTED to launch his audacious winter counteroffensive, known as Operation Watch on the Rhine,
on November 27, 1944, the time when cold fog typically begins
to blanket the Ardennes, but his plan was delayed for several
weeks—primarily due to Nazi fuel shortages—and only ready by
mid-December. The führer wanted to personally direct the massive operation from his western headquarters, located in the rural
Taunus Mountains in central Germany.

Hitler traveled by armored train to Giessen on the blustery
morning of December 11, 1944, then was chauffeured the short
drive to Ziegenberg Castle in his black Mercedes sedan. Unknown
to Allied military intelligence at the time, or even to the local
townspeople, a compound was disguised behind the imposing
castle, built in the mid-1700s; it was "made up of seven buildings
giving the appearance of an innocent grouping of wooden country cottages with second-story dormers. Many even had wooden
porches decorated with flower baskets. They were bunkers and

had 3-foot thick walls and ceilings of reinforced concrete." The name Hitler had given to this disguised and fortified compound was Adlerhorst—Eagle's Eyrie.

Hitler's plans had been drawn up in the greatest secrecy, with him obsessively controlling every detail. Ultimately, all generals had been forced to sign an oath of secrecy, which specified a death penalty for leaks. Late that December afternoon, buses brought divisional commanders to Adlerhorst to be personally briefed by Hitler about the counteroffensive. Each officer was first searched by SS men and ordered to surrender both his sidearm and his briefcase.

The führer's paranoia was full-blown: the failed assassination attempt of July 20, 1944, had left him twitching, his left arm palsied, his face swollen and puffy. In the wake of the devastating loss at Stalingrad and the successful Allied invasion at Normandy, Colonel Klaus von Stauffenberg and the other plotters had aimed to assassinate Hitler and wrest control of the armed forces from the Nazi Party and the SS so a separate peace could be made with Britain and the US as soon as possible. The failure of the assassination and the military coup led to 7,000 arrests by the Gestapo—with 4,980 immediately executed. Afterward, Hitler no longer trusted his Prussian generals or most of the Wehrmacht High Command. Instead he put his faith in the fanatical SS.

Hitler's behavior seemed to many of the generals increasingly erratic; he took a daily cocktail of cocaine and amphetamines, administered intravenously by his quack personal physician. The generals who had not seen him since the assassination attempt were also taken aback by Hitler's physical appearance: he was pale and stooped over, and his left arm trembled uncontrollably.

At 1800 hours the führer sat behind a table and launched into a long rambling diatribe about Frederick the Great, Hitler's own visionary role in the early Nazi conquests of 1939 and

1940—reminding his generals of their skepticism of his strategic genius—and a justification that this war was one of self-preservation. Finally, he announced the details of the attack: at 0530 on December 16 three armies would punch a hole in the thinly defended Allied lines in the Schnee Eifel. Most of the commanders could scarcely believe that Hitler had managed to amass 410,000 troops, 1,400 panzers, and 2,600 artillery pieces for the predawn offensive.

It would be a return to the glorious victories of the 1940 blitzkrieg, Hitler predicted. This last gambit would reverse the Reich's string of bad military losses. Using the element of surprise and the bad weather—which would neutralize Allied air superiority—his divisions would race through the forest, cross the bridges of the River Meuse, and press onward until they seized the port of Antwerp. In the process, Operation Watch on the Rhine would split British and American forces in northern France. If the foul weather held and the panzers and infantry could keep to their timetable, the Allies would be caught completely off guard.

"This battle is to decide whether we shall live or die," Hitler told the assembled generals. "I want all my soldiers to fight hard and without pity. The battle must be fought with brutality and all resistance must be broken in a wave of terror. In this most serious hour of the Fatherland, I expect every one of my soldiers to be courageous and again courageous. The enemy must be beaten—now! Thus lives our Germany!" Hitler added: "The enemy can never reckon upon us surrendering. Never! Never!"

The vanguard of the attack would be the Sixth Panzer Army, led by SS-Oberstgruppenführer Joseph "Sepp" Dietrich, one of Hitler's closest confidants going back to the 1920s. Dietrich had been a sergeant in World War I, a butcher, a street brawler with the Brownshirts during the Nazi Party's rise to power, and one of the rabble who'd fought alongside Hitler in the Munich Beer Hall

Putsch. The special protection unit that Dietrich founded in 1932 had evolved into the Leibstandarte SS Adolf Hitler, which served as Hitler's personal army and later its own militarized division in the Waffen-SS.

Dietrich's rough manner and storm-trooper background earned him little respect from the old-line Prussian generals. He was said to be unable to read maps, and one senior Wehrmacht commander derided him as a "conceited, reckless military leader with the knowledge and ability of a good sergeant." Nonetheless, Dietrich had Hitler's full confidence, and it was his Sixth Panzer Army that would spearhead the counteroffensive. Dietrich was to attack the Allies' northern flank in Belgium—from Monschau to the Losheim Gap—roll over Elsenborn Ridge, race for the Meuse River, and on to Antwerp.

If Dietrich lacked the respect of seasoned Wehrmacht generals, the same could not be said of the commander of the Fifth Panzer Army. General Hasso von Manteuffel, born into a family of aristocratic Prussian military officers, stood just over five feet tall, was slightly built, and had been an excellent horseman and jockey as well as a decorated combat officer in World War I. Despite his diminutive stature, Manteuffel was known as tough-minded and, like Dietrich, unafraid to rebut Hitler and offer his unadorned criticism. When the Ardennes offensive launched, with two of his divisions, Manteuffel had orders to encircle the Schnee Eifel salient, trapping the American 106th Infantry; with his remaining forces, he was told to take the towns of Saint Vith and Bastogne, the most important rail and road junctions in the region, and then advance northwest to protect Dietrich's southern flank. Meanwhile, General Erich Brandenberger, commanding the Seventh Army, comprised largely of infantry, would attack even farther south and block any Allied attempt to send in reinforcements.

This would be a conflict unlike any seen on the Western Front, largely because Hitler had given so much responsibility to his trusted SS, many of whom were by now veterans of the savage killing fields of the Eastern Front. No single commander characterized this better than SS-Obersturmbannführer Joachim Peiper. Twenty-nine years old, Peiper had striking good looks, with bright blue eyes and brown hair perfectly slicked back. Even by the standards of the SS, he was known for being particularly fanatical and ruthless. He had enlisted in the SS at nineteen and was soon assigned to Dietrich's elite Leibstandarte SS Adolf Hitler.

Peiper's training in the SS had been at Dachau—the Reich's first concentration camp—and by July 1938 he was working directly for Heinrich Himmler. In June 1941, at the age of twenty-six, Peiper participated in Operation Barbarossa, the Nazi invasion of the Soviet Union. He soon became Hauptsturmführer—or captain—in an SS combat unit, and between 1941 and 1943 he led several units in the Soviet Union, plus in Italy, where he was responsible for the massacre of no fewer than twenty-two civilians. In the Soviet Union he was infamous for encircling and burning down entire villages, killing all the inhabitants, earning his SS unit the nickname Blowtorch Battalion. Often Peiper and his men simply machine-gunned Russian soldiers rather than take any prisoners of war. They were especially ruthless against partisans they fought in the eastern woods, summarily executing all they encountered.

In the days ahead, as commander of his Kampfgruppe Peiper—or Fighting Group Peiper—he and his troops would commit some of the most ruthless and infamous acts in the Second World War.

HITLER HAD ALSO made sure that in his last audacious military gamble there would be a role for his favorite SS commando, Austrian Otto Skorzeny.

Skorzeny was an imposing figure: six foot four, powerfully built, and sporting deep scars on his left cheek from a fencing duel. By December 1944, he had already achieved a mythical status within the German military. Rank-and-file SS men "almost worshipped Skorzeny as a super-hero" after his numerous commando missions, including the rescue of Benito Mussolini in a daring airborne raid from his imprisonment by the Italian Resistance in the Gran Sasso mountains.

"Skorzeny, this next assignment will be the most important of your life," Hitler had told him on October 21 when he summoned Skorzeny to the Wolf's Lair headquarters in East Prussia. Hitler then explained that Skorzeny was the only man he could entrust with the success of Operation Greif—Operation Griffin—the personal brainchild of the führer. The ultimate purpose of this top-secret operation would be to capture the strategically vital bridges over the Meuse River and to create havoc behind enemy lines.

> I want you to command a group of American troops and get them across the Meuse and seize one of the bridges. Not, my dear Skorzeny, real Americans. I want you to create special units wearing American uniforms. They will travel in captured Allied tanks. Think of the confusion you could cause! I envisage a whole string of false orders which will upset communications and attack morale.

Hitler granted Skorzeny virtually unlimited authority to prepare his mission. Using tactics entirely illegal under international laws of warfare, his soldiers would dress in authentic US Army uniforms, drive captured jeeps and Sherman tanks, and operate as undercover saboteurs behind the Allied lines during the Ardennes offensive.

The first piece of business was to find men who could talk with plausible American accents. According to historian Antony

Beevor, "Officers and NCOs from the army, Waffen-SS, Kriegs-marine, and Luftwaffe who spoke English were ordered to report to the camp at Schloss Friedenthal outside Oranienburg for 'inter-preter duties.'" They were interrogated in English by SS officers, then told they would be part of a special unit called the 150th Pan-zer Brigade and ordered to sign a secrecy order, any violation of which would be punishable by death.

A young naval officer named Muntz was given the task of col-lecting more than two thousand US military uniforms, of officers, NCOs, and enlisted men, from prisoner-of-war camps by late No-vember.

For his commandos, Skorzeny selected 150 men out of the more than 600 troops fluent in English. All the English speakers were housed in a special camp, wore GI uniforms around the clock, and were fed captured K rations. Every order was given in En-glish, and they learned to salute in the American style. They spent weeks playing the role of US soldiers, with training that included watching Hollywood movies to learn up-to-date slang and even learning "how to tap their cigarette against the pack in an Amer-ican way." Anyone who questioned the mission was threatened with summary execution.

After penetrating the lines during the Ardennes counterof-fensive, the fake Americans of the 150th Panzer Brigade would set out to destroy bridges, ammunition dumps, and fuel stores, send bogus orders to any US units they met, reverse road signs, remove minefield warnings, and cordon off roads with warnings of nonexistent mines. Plus, they would disrupt the US chain of command by destroying field telephone wires and radio stations and issuing false orders.

BY DECEMBER 14, on the German side of the front, all the final preparations for Operation Watch on the Rhine were going into effect. Corduroy roads to the front line were laid, covered with thick beds of straw in an attempt to muffle the sounds of the panzers and half-tracks. Security was tightened in towns and villages that bordered the Ardennes.

How all this massive movement of troops—hundreds of thousands of men, tanks, trucks, and artillery—had been kept secret until the middle of December remains a subject of debate among historians. Without doubt, it was one of the most massive military intelligence failures of the entire war.

General Kenneth Strong, working in SHAEF's G-2 intelligence section, concluded an attack was impending, but when he brought his concerns to the 12th Army Group's intelligence section, they were discounted. SHAEF assumed that a buildup of German troops and weaponry was most likely defensive.

Colonel Benjamin A. "Monk" Dickson, the chief intelligence officer for the US First Army Group, had also tried to sound warning alarms with his superiors. On December 10, 1944, in a report titled "G-2 Estimate Number 27," Dickson wrote: "It is plain that his [Hitler's] strategy in defense of the Reich is based on the exhaustion of our offensive to be followed by an all-out counterattack with armor, between the Roer and the Erft, supported by every weapon he can bring up to bear." Dickson further noted that the morale among German prisoners of war had also "achieved a new high" and some were even attempting escapes in efforts to rejoin Wehrmacht forces. For Dickson, all this evidence contradicted 12th Army Group's "assessment that the German army was a beaten and vanquished foe." A civilian named Elise Dele had been picked up by a German patrol, escaped, and was smuggled by members of the Resistance across the lines into Luxembourg, where she offered an eyewitness account to intelligence officers of the 28th Infantry Division of having seen "bridging equipment on the German side of the Our River"—equipment that could only have been preparatory for the advance of panzers, artillery, and other vehicles. Her report moved at a snail's pace through the US Army bureaucracy, eventually reaching Colonel Dickson at First Army Group. "It's the Ardennes!" Dickson is reported to have said to First Army's commander, General Courtney Hodges.

But Dickson's repeated warnings were ignored.

Even among the German High Command there had been leaks of the buildup in the Ardennes. Hitler's order for total secrecy was not strictly followed, asserts Antony Beevor. "Word of the forthcoming offensive even circulated among senior German officers in British prisoner-of-war camps." As early as the second week of November, "General der Panzertruppe Eberbach was secretly recorded saying that a Generalmajor Eberding, captured just a few days before, had spoken of a forthcoming offensive in the west with forty-six divisions. Eberbach believed this was true and that it was a last try." Another Nazi lieutenant had heard that "the big offensive, for which they were preparing forty-six divisions, was to start in November." These secret conversations were reported by British Intelligence to the War Office in London and sent on to SHAEF, but this "vital information does not appear to have been taken seriously."

The problem was, of course, that even if the intelligence sources were solid, the very idea behind a massive wintertime offensive in the west seemed too ludicrous, too far-fetched, to be true. "No goddamn fool would do it," said Lieutenant General Walter Bedell Smith, Eisenhower's chief of staff. At the most, Allied intelligence thought there might be a diversionary attack to relieve pressure elsewhere on the Siegfried Line.

The Allied commanders simply couldn't believe the Germans would attempt to go on the offensive in the west when they clearly "needed to husband their strength before the Red Army launched its own winter onslaught." Such a gamble was not in keeping with the conservative style of Field Marshal Gerd von Rundstedt, "but the Allied command had gravely underestimated Hitler's manic grasp on the levers of military power."

Senior US officers also believed that because they wouldn't attack under such unfavorable conditions, the Nazis wouldn't either. This failure of intelligence gathering, strategic understanding, and the basic psychology of Hitler would prove tragic by mid-December 1944.

TWELVE

A T 0530 HOURS on December 16, 1944, the frozen earth erupted: hell appeared like a ghost in the forest.

In an instant, pine trees exploded into deadly wooden spikes. The frigid air turned fiery red. Blood and bone mingled with chunks of thawing debris.

Roddie clung to the shaking icy ground, desperately trying to crawl into his helmet. Terrain and weather were no longer his greatest enemies. His enemies were the relentless concussions and deadly shrapnel from the murderous 88s.

The German artillery rained down with pinpoint accuracy from what seemed like every direction. No one was safe. There was nowhere to run or hide—just across the valley were thousands of enemy troops, with thundering panzers and heavy artillery.

Fear was also Roddie's enemy. Fear that he would be blown to pieces in an instant. Fear for the lives of his boys—Lester, Frankie,

Skip—and what they were experiencing. Fear that he might not make it home, might never see his family again.

This was a terror Roddie never could have imagined. He knew panic was lethal to an infantryman, but it gripped him with a sickening embrace. His body shivered as he shook off thoughts of his own death.

The treetop calibration tactic perfected in the Hürtgen Forest was being put to devastating effect now in the Ardennes. Every tree seemed to have been simultaneously blasted from its roots.

Even with all the live-fire training back at Fort Jackson and Camp Atterbury, there was no way Roddie, or anyone, could have prepared for the horrific reality of battle—especially *this* battle. Roddie fought an overwhelming urge to run. All around him wounded men were screaming. It was suddenly hellfire in their "quiet" sector of the Ardennes, and Roddie realized that the image of a triumphant march through Berlin by Christmastime had been ludicrous.

Though fiery shrapnel continued to rain down, Roddie overcame the urge to panic and crawled toward the safety of the sandbags enclosing the communications station for Headquarters Company of the 422nd Regiment. There, Roddie reached for his radio. Dazed, he called Regimental Command at Schlausenbach to relay the intensity of the attack, the frightening accuracy of those Nazi 88s.

Roddie's commanding officers downplayed the incoming reports as an overreaction since, unlike the 28th, which had fought through the hell of Hürtgen, the 106th had yet to see serious combat.

His dugout was claustrophobic. During a momentary break in the shelling, he emerged for a breath of fresh air only to hear a whistle—a bullet fired from an unseen Maschinengewehr 08, flashing inches above his head. In his diary he described the near-fatal moment:

Boy, I was thankful the Lord was on my side. And I didn't hesitate to tell him either. I prayed hot and heavy and was convinced of that old saying—that there are no atheists in a foxhole.

In the furious chaos of artillery fire all around, some GIs were saying that it felt like the end of the world had come.

"Our little village of Schlausenbach was hit time and time again," Lester recalled. "The houses no longer looked secure, and we were digging ourselves into the foundations and into the ground." GIs who had been trained as regimental headquarters personnel now became riflemen to repulse attack after attack. "I can still see Butch, our Headquarters Company cook, dashing around in the jeep and firing his machine gun like mad," Lester wrote. "None of us knew how serious the situation had become."

The dreaded 88 mm shells "bored through the darkness at half a mile per second, as if hugging the Ardennes hills," and the Nebelwerfer rockets—veteran GIs had dubbed them "Screaming Meemies"—resounded "in the hollows where wide-eyed GIs crouched. Then enemy machine guns added their racket to the din, and rounds with the heft of railroad spikes splintered tree limbs and soldiers' bones alike."

Then Roddie and the infantrymen heard the low, almost primeval groan of panzers and creaking armored vehicles as the thrust of Hitler's last gamble burst forward through the forest. The German infantrymen appeared camouflaged in white snowsuits, some in Wehrmacht greatcoats, screaming like banshees. The Nazi artillery gunners seemed to know the American positions down to the millimeter—the GIs were, after all, fighting from the Germans' own defensive bunkers.

"Strange lights lit up the sky," recalled Sergeant John Morse. "German searchlights aimed at the low clouds reflected down to the ground."

In the predawn hours, two GIs were "idly chatting in front of an abandoned Siegfried Line bunker." The men were smoking cigarettes, talking about their families back in the States, "complaining about the bitter cold and having to pull lookout duty at such an ungodly hour of the morning."

One of the GIs was stunned by the sight of the massive searchlights eerily "piercing the misty gloom illuminating heavy clouds above their heads. 'What the hell is that?' he asked. Suddenly there was a whizzing sound, followed by a blinding flash of light as he exploded into thousands of pieces. A mortar shell had impacted his body and in a split second reduced it to unidentifiable flying chunks of bloody flesh that slapped down on the cold hard snow like fresh meat on a butcher's block. The other GI dived for cover, pressing one hand on his helmet and muttering inanely as he burrowed frantically, shells bursting around him."

Roddie soon found that the Nazis were jamming all radio frequencies and had severed telephone wires, making transmissions impossible. Later he would grimly note in his diary that the Germans had done "a pretty complete job" destroying the communications.

The unrelenting artillery and mortar fire continued to rain down on the GIs—and on positions to their rear—for close to an hour. Then the German soldiers advanced across the open fields in a column of what appeared to be platoons in line, with some elements working their way up to the wooded ridges. US riflemen and machine gunners opened fire. Mortar fire was called for, too, and was aimed at the ridges. By midmorning, the main thrust of the German attack was stopped, though isolated patrols of Germans were still infiltrating through B Company's lines. One of the patrols well camouflaged in white snowsuits succeeded in getting within approximately a hundred yards of the command post before they were either killed or taken prisoner.

The German prisoners were immediately sent to Battalion Headquarters for interrogation. One captured POW was a Wehrmacht officer who had on him a written copy of the attack order. A corporal in Battalion Headquarters, fluent in German, quickly translated the order, and the officers were stunned to learn that this was no minor patrol activity but indeed part of a massive German counteroffensive. The regimental headquarters in Saint Vith was immediately notified by special motor messenger.

DESPITE THE DOGGED resistance of the 422nd and 423rd, the two forward regiments of the 106th Division, Manteuffel's Fifth Panzer Army came down "like wolves on a sheepfold, falling on an American regiment at an unnerving ratio of ten wolves for each sheep." Meanwhile, the Waffen-SS exploited the overstretched US defensive positions, opening up a six-mile-wide gap between the 106th and 99th Divisions. In fact, powerful spearheads of tanks and infantry, to both the left and the right, were in the process of trying to envelop the men of the 106th in a pincer move around the main ridge of the Schnee Eifel.

"On no segment of the Western Front were GIs more outnumbered, yet sharp firefights that morning imperiled the German timetable," according to historian Rick Atkinson.

To Roddie, all the infantrymen under his command were putting up heroic resistance; the former college students, the ASTPers, may have been green, but now, in the fury of the conflict, they were transformed from teenagers into men, battling the Germans with brains as well as brawn. Roddie was proud of all of his boys; they were brave in the face of death.

Much like Major General Jones's 106th Division, Major General Norman Cota's 28th Infantry Division was tasked with holding an impossibly wide front—some twenty-five-miles—along the Our River, with its three infantry regiments on the line. Instead of facing two German divisions across the river, as army intelligence had reported, Cota's men found themselves fighting five full divisions, plus heavy enemy reinforcements.

The German artillerymen had excellent intelligence on the US troops. They knew the GIs preferred to sleep in the homes of villagers, rather than in foxholes, so many of the initial artillery bursts fell on civilian houses.

Sonny Fox found that out firsthand. He was with Cota's 28th Division, bunking in farmhouses and barns. When the German fusillades started on the morning of December 16, 1944, Sonny was awakened at 0520 by artillery screeching over the farmhouse. He ran to a foxhole a few yards from the building and jumped in to find his sentries.

"How long has this been going on?"

"About twenty minutes," the sentries said.

Sonny started back to the house to get the rest of his squad out. "As I got close to the door," he recalled, "I heard an incoming and dove under the butcher block alongside the entrance. The shell passed over and I started to extricate myself but could not get free. The battle was beginning, and I was stuck under a butcher block! I started to giggle. All hell was breaking loose and I was going to end my life stuck under a butcher block. I finally broke loose and dashed into the house to find my mighty band of warriors cowering in the potato cellar. I shouted them out of there and into our prepared positions on both sides of the farmhouse. Unfortunately, that meant I was out of contact with those on the far side of the building. I was with two of my squad; the others were scattered about."

And then, as suddenly as it had begun, the artillery barrage stopped. Eerie green signal lights arched into the sky against the thick cloud cover.

Sonny recalled that "at one point, I heard some Germans walking down the road just on the other side of the hedge that separated us from that road. As they approached our position, I pulled the pin from a grenade. I rose up enough to catch a glimpse of three soldiers chatting as they walked, as though they were on a stroll down the Unter den Linden [a boulevard in central Berlin]. I tossed the grenade and watched long enough to register their startled reaction. Then I slid back into the foxhole to avoid the explosion. There is a pause of 4.5 seconds before the grenade explodes. It's amazing what I was able to see in those seconds. One of the men was wearing a Red Cross armband. I did not know this before I threw the grenade."

All three Germans were extremely young—Sonny estimated they were around his age, eighteen or nineteen years old. The soldier with the Red Cross armband had on dark-rimmed glasses. Sonny would forever remember the terror in the young men's eyes, just before all three Germans were blown to pieces.

By the afternoon of that day, Sonny heard tanks coming down the road. He was not sure if they were German or American. "They turned out to be ours, and I thought we had beaten them off, but the captain said no, the Germans were behind us and the fighting was going on well behind our lines."

Indeed, huge numbers of German troops and tanks were well to the rear of the Allied lines; some were also masquerading as US soldiers behind the front. Otto Skorzeny's disguised commandos of Operation Greif were wreaking havoc with Allied communications. The brigade, dressed in their authentic US infantry uniforms, driving captured jeeps and Sherman tanks, toting American weapons, had managed to penetrate the Allied lines during the early hours of the Ardennes counteroffensive and were carrying out their planned disruptions. One commando team persuaded a US Army unit to withdraw from the Belgian town of Poteau, northwest of Saint Vith, and by switching around road signs, another team sent an entire US regiment in the wrong direction.

But word soon got out that German commandos, speaking good English and dressed in American uniforms, were behind the lines, and the US Army hastily set up checkpoints. GIs began grilling everyone—even high-ranking officers—on facts they felt only real Americans would know: minute details about baseball, US geography, and Dick Tracy's villains. "What team is known as Dem Bums?" "Where does Li'l Abner live?" "Who's Flattop? Pruneface? The Mole?"

While they succeeded in capturing a few of the German commandos, the roadblocks often failed miserably. American soldiers shot out the tires on British Field Marshal Montgomery's jeep after he refused to stop at a bridge checkpoint. Brigadier General Bruce Clarke was also detained at gunpoint for five hours after he incorrectly told military policemen the Chicago Cubs played in the American League.

"But I'm General Bruce Clarke!" he shouted.

"Like hell!" an MP snapped back. "You're one of Skorzeny's men. We were told to watch out for a Kraut posing as a one-star general."

Even Clarke's commander, General Omar Bradley, was repeatedly stopped at checkpoints. "Three times I was ordered to prove my identity by cautious GIs," Bradley recalled. "The first time by identifying Springfield as the capital of Illinois (my questioner held out for Chicago); the second time by locating the guard between the center and tackle on a line of scrimmage; the third time by naming the then current spouse of a blonde named Betty Grable. Grable stopped me, but the sentry did not. Pleased at having stumped me, he nevertheless passed me on."

Paranoia spread quickly. All over the Ardennes US soldiers attempted to persuade suspicious guards and MPs that they were genuine GIs. One infantry captain was held because he was observed to be wearing German boots.

Most of Skorzeny's fake Americans were apprehended in the dragnet, but so great was the confusion caused by Operation Greif that the US Army began to see spies, saboteurs, and assassins everywhere. "A half million GIs played cat and mouse with each other each time they met on the road," Bradley remembered. "Neither rank nor credentials spared the traveler an inquisition at each intersection he passed." Two US soldiers were even reported to have been shot by a trigger-happy US MP who was convinced they were Skorzeny's SS men in disguise.

The captured commandos offered up clever disinformation under questioning. One claimed that Operation Greif's actual mission was to have Skorzeny—who was still at large, dressed as an American officer, somewhere behind the lines—assassinate General Eisenhower. Another said Ike and his entire staff were going to be kidnapped by Skorzeny and held as negotiating hostages. Far-fetched stories—but they did result in Eisenhower being confined to SHAEF headquarters, under close guard, for his own safety.*

EVERYWHERE BEHIND THE front lines, confusion reigned. By midday on the sixteenth, Major General Jones, commander of the 106th, had thrown virtually all the divisional reserves into combat.

The intelligence from frontline reports coming into Jones

---

* Eisenhower was unamused by having to spend Christmas 1944 isolated and under close guard in the Trianon Palace Hotel in Versailles for his own protection. After several days of confinement, Ike left his office, angrily declaring he had to get out and that he didn't care if anyone tried to kill him.

was limited, due to the disruption in communication lines, but he knew the Nazis had flooded through the Losheim Gap and were attempting to encircle his entire division on the Schnee Eifel. What he didn't yet know is that the 14th Cavalry Group, assigned to protect the northern flank of the 106th Division, had been nearly decimated, in retreat, and the Germans were marking Saint Vith itself as the next day's main objective.

Without a full understanding of the military calamity unfolding, Jones scrambled for a strategy. Should he order the 106th's three regiments, dug in along the twenty-eight-mile front, to pull back? Or should he wait? From a command post located in an old schoolhouse in Saint Vith, Jones, lacking combat experience, decided to ask advice from the VIII Corps commander, Major General Troy Middleton.

"You know how things are up there better than I do," Middleton said in a crackling phone call from his own headquarters in Bastogne. "But I agree it would be wise to withdraw them."

A brief disruption in the telephone line seems to have affected the entire course of the battle. Jones never heard the second sentence. He hung up, telling his staff in Saint Vith, "Middleton says we should leave them in."

At that same moment, General Middleton told his subordinate officer in Bastogne, "I just talked to Jones. I told him to pull his regiments off the Schnee Eifel."

That misunderstood communication would decide the fates of the 422nd and 423rd Regiments. Roddie and soldiers of the Golden Lions would hold their ground, despite the howling enemy swarming past them on both flanks.

It was a catastrophic strategic mistake; some military historians have argued that, in his defense, Jones was under the impression that thousands of reinforcements were on the way—mere hours away.

But behind the front lines, the roads were jammed with traffic fleeing in the opposite direction. There was utter confusion and chaos. "It was a case of every dog for himself," one major recalled. "The most perfect traffic jam I have ever seen." A tanker driving a Sherman, trying to plow forward against the westward exodus, reported that "the fear-crazed occupants of the vehicles fleeing to the rear had lost all reason."

Jones's decision not to pull back his men left not just the two infantry regiments of the 422nd and the 423rd completely exposed to entrapment on the Schnee Eifel but also five artillery battalions. Plus, since the Nazis had wiped out all communications lines, the men had no air support or tanks. They never stood a chance. As Roddie later wrote, what good were rifles against tanks and 88s?

Despite overwhelming odds and running out of ammunition, the brave boys of the 106th fought tenaciously to stop the German advance. If they couldn't push the enemy back, at the very least maybe they could slow them, keep them from breaking out and achieving their objective of taking the port of Antwerp.

One thing was certain: the men of the 106th's forward regiments weren't taking a backward step.

In the heat of a furious exchange of gunfire, a German soldier shouted sarcastically, in English, "Take a ten-minute break. We'll be back."

A GI, hidden in the frozen forest, could be heard shouting back, "We'll still be here, you son of a bitch!"

SKIP FRIEDMAN HAD pulled guard duty that first day of the attack. He was already suffering the effects of the snow and damp—as he marched his post, his combat boots were leaking, and he stomped his feet constantly to keep his toes from going numb. Only a few months earlier he'd been in uniform on the University of Alabama

campus at Tuscaloosa, in ranks with the other Quiz Kids, studying thermodynamics, calculus, and mechanical engineering. Now
he found himself trying to fend off frostbite and trench foot, and
being blown to bits by the rain of Nazi 88s.

On the night of the sixteenth, Skip entered the headquarters of
the regimental commander, Colonel George L. Descheneaux Jr.,
and saw a detailed map hanging on the wall. "There are arrows
north of us, south of us, and *behind* us," Skip said. "I'm a T-Five at
that time—meaning, I'm just a corporal. Even I—as a corporal—
can look at that map and say, 'That is not good.' And it was not
good. We were in this little town of Schlausenbach and we realized that our position was untenable."

THIRTEEN

BY THE FOLLOWING DAY, Sunday, December 17, the German trap was snapping on Roddie and the two forward regiments of the 106th Division. The panzer columns and infantry from north and south converged in their pincer movement at the town of Schönberg, to the east of Saint Vith. By dusk, "nine thousand GIs were surrounded on a bleak, snowy German moor. An icy west wind whipped through the fir trees, carrying the fateful whine of panzer engines."

The men of the 422nd didn't realize yet the dire position they were in. "None of us knew how serious the situation had become," Lester recalled. "I discovered it quite by accident on the night of the seventeenth when I walked into the command post and found the clerks quietly burning papers and documents. The next day I was told to burn all our communication codes, that we were slipping out of town to try to break through to our lines. That was [the] first indication I had that we were surrounded and cut off from our supplies."

At first light, a few miles to the north, the atrocities began. Shortly before 0600, the Kampfgruppe Peiper rolled into the town of Honsfeld to find several American vehicles parked with exhausted GIs sleeping inside.

Joachim Peiper, veteran of those Russian massacres, the man responsible for blowtorching women and children alive, would now show his true colors here in Belgium.

Eight American soldiers were rousted outside in their underwear and bare feet, shouting, "I surrender!" Peiper's Waffen-SS men lined the defenseless GIs up in the street and mowed them down with a machine gun. Five others emerged from a house, waving a white flag; four were shot, and the fifth, pleading for mercy, was crushed to death beneath a tank. Four more Americans, also carrying a large white banner, were shot. Peiper's men stripped boots from the dead and continued on in their panzers to Büllingen, two miles northwest.

MEANWHILE, THE ALL-BLACK troops of the 333rd Field Artillery Battalion—equipped with 155 mm howitzers—were just behind the front lines, on the Andler–Schönberg road, east of Saint Vith, fighting in support of the 106th Division. Like all segregated units in World War II, white officers commanded black troops, but unlike the inexperienced men of the 106th, the 333rd consisted of combat veterans who prided themselves on being able to pick off German tanks at great distances with their accurate howitzers.

Among them was Staff Sergeant Sam Harris, a young artillery gunner from rural Rowan County, North Carolina. The youngest of five siblings, Sam had worked briefly in an aircraft factory after finishing high school, then enlisted in the US Army on July 16, 1942, at the age of eighteen, and was inducted at Fort Bragg. With the 333rd, Sam had landed on the beaches of Normandy in July 1944 and had seen some of the most vicious and continuous combat of any US artillery outfit throughout the summer.

At the onset of the Battle of the Bulge, Sam and his outfit were roughly eleven miles behind the front. By the morning of December 17, the Germans had captured Schönberg and controlled the bridge across the river that connected to Saint Vith. Many of the 333rd Field Artillery were killed; most of the rest of the battalion were captured. As the captured men were being herded to the rear, the column was strafed by friendly fire—attacked by a US aircraft. During the ensuing confusion, eleven black soldiers managed to escape into the snowy woods.

The eleven men tried to find the American lines but became badly lost; they trudged through waist-deep snow, staying away from roads, hoping to avoid German patrols. Between them they carried only two weapons. Exhausted and hungry, the men stumbled upon the tiny Belgian farming village of Wereth shortly before dusk.

They were waving a white flag.

A friendly farmer named Mathias Langer offered them food and shelter. Like many tiny towns in the region, Wereth was deeply divided in its loyalties, and the wife of a German soldier quickly notified a four-man patrol of the First Waffen-SS Division about the eleven black American GIs hiding at the Langer farm.

The men hadn't finished eating when an SS vehicle pulled up to the Langer farmhouse. The eleven emerged, hands up. The SS ordered them down on the damp ground behind the house.

It was growing dark; the men were shivering. Defenseless and compliant, the black soldiers were taken to a nearby field, where they were slowly tortured, maimed, and shot. They had their eyes gouged out by Nazi bayonets—while still breathing. They were stabbed and rifle-butted, and had their fingers severed. The SS men then rolled their armored cars over them. They were finally shot in a sadistic fashion, meant to inflict maximum suffering, not death. One black soldier was shot while in the process of bandaging a comrade's wounds.

News of the savage torture-killings of the "Wereth 11" would not come to light immediately. But events that transpired several miles away, that same day, would quickly become known as the worst atrocity committed by Nazi troops against US forces during World War II.

Just between noon and 1 p.m., Joachim Peiper's SS panzer regiment approached the Baugnez crossroads, two miles southeast of the town of Malmedy. An American convoy of about thirty vehicles, mainly from the 285th Field Artillery Observation Battalion, had the misfortune to arrive at the crossroads at the same time, turning toward Saint Vith, where it had been ordered to join the Seventh Armored Division in defense of the city.

Peiper's panzers opened fire, hitting the first and last trucks in the convoy, setting them ablaze. Like Roddie and his men, these Americans were armed only with rifles and small arms. They tumbled off the trucks, trying "to seek shelter or run for the forest. Panzergrenadiers rounded up some 130 prisoners and herded them into a field by the road."

The armored column led by Peiper continued west toward Ligneuville. The SS troops left behind barked at the GIs and took "rings, cigarettes, watches and gloves from their prisoners."

At 1415, the unarmed US soldiers were standing in a snowy field near the Café Bodarwé when, without warning, and for reasons that are still unclear today, the SS troops began mowing them down with machine guns.

As soon as the Nazis opened fire, the prisoners panicked. Some tried to flee, but most were murdered right where they stood. Others feigned death, hoping to be spared in the confusion, but the SS men walked among the bodies and shot anyone who appeared to be alive in the back of the head with their Lugers. A few other American soldiers looked for shelter in the café at the crossroads; the SS men surrounded the café, set fire to the building, and shot any American who tried to escape.

Though the mass murder occurred at the Baugnez crossroads, the killing of what amounted to eighty-four defenseless prisoners of war was to become infamous as the Malmedy Massacre.

Malmedy. Within hours that melodic Walloon name would spread terror all along the front, back to SHAEF headquarters, striking fear and rage in all American hearts.

Though Peiper was not present for the atrocity, it was clear his men had followed Hitler's command to engage the Americans with all the brutality they'd previously exhibited against the Soviet Army on the Eastern Front. Germany was now in what Joseph Goebbels had called a state of *totaler krieg*—total war. The Geneva and Hague regulations on the humane treatment of captured enemy combatants had been utterly discarded.

Around 1430 that afternoon the first survivors of the massacre were found by a patrol from the 291st Engineer Combat Battalion. Some came out of hiding shortly after the SS departed and returned through the lines to nearby Malmedy, where US troops still held the town. Eventually, forty-three survivors emerged.

The inspector general of the First Army learned of the shootings about three or four hours later. One military policeman who'd witnessed the whole incident was brought to First Army headquarters in the town of Spa. He described to General Courtney Hodges and his other senior officers how the prisoners had been "herded together into a side field and an SS officer fired two shots from his pistol and immediately there came the crackle of machine gun fire and whole groups were mown down in cold blood."

Rumors that the enemy was killing prisoners soon reached the forward American divisions. Shock and fury were widespread. "Immediate publicity is being given to the story," General Hodges's chief of staff noted. By evening, the news had reached General Eisenhower at SHAEF in Reims and General Bradley in the 12th Army Group headquarters in Luxembourg. Details of the Malmedy Massacre had spread like wildfire to all the US military posts on the front.

Allied High Command issued an urgent warning to all troops: "It is dangerous at any time to surrender to German tank crews, and especially so to tanks unaccompanied by infantry, or to surrender to any units making a rapid advance. These units have few means for handling prisoners and a solution used is merely to kill the prisoners."

Even German generals being held POW in England were horrified and visibly shaken when the news reached them about the Malmedy Massacre. "What utter madness to shoot down defenseless men?" one said. "All it means is that the Americans will take reprisals on our boys." Another added, "Of course, the SS and paratroopers are simply crazy—they just won't listen to reason."

RODDIE AND HIS BOYS continued to battle furiously around Schlausenbach. Like most of the 106th Division, they heard some of the graphic accounts from eyewitnesses and survivors of Malmedy. The news changed everything. After Malmedy, the GIs knew that any US soldier—regardless of race or religion—who tried to surrender his weapons was liable to be shot by the Nazis in cold blood. The news also strengthened the will of the GIs, their blood boiling with righteous indignation at the cruelty of the Nazis.

Though circumstances were growing ever more desperate, since the men were running out of ammunition, fuel, and food, they vowed to fight on.

That night, Roddie wrote in his diary:

I enjoyed my last meal on the evening of the 17th, because the morning of

FOURTEEN

J UST AFTER 11 A.M. on December 18, in the tiny town of Hosin-
gen, Luxembourg, Sonny Fox and the men of his unit of the
110th Regiment, 28th Infantry Division, surrendered, trudging
out of the houses they still held in the town. "We hadn't taken
many casualties in our own unit and had destroyed our jeeps and
our weapons," Sonny recalled. "Out we came, with our hands
raised high. My emotions were wildly mixed. Relief at being alive,
fear of what was about to befall us, but also a keen curiosity about
what this new experience would be like. I was curious to walk
through the looking glass and see the German side of the war. I
think my concerns were mitigated by being part of an organized
surrender. I was still with my unit. I was still an American soldier. The scene just beyond our houses underscored the wisdom
of the captain's decision. The Germans had brought up tanks and
artillery, and they were all aimed at the few houses we had been
holding. They were about to level us."

Sonny and the other infantrymen in his outfit were marched

from the field back to the original German front: the border between Luxembourg and Germany at the Our River. As they marched past the twisted bodies of dead German soldiers, Sonny recalled, "one of the guys in my squad piped up, in a fairly loud voice, 'Jeez, we got a lot of 'em, didn't we?' as we were marching *under guard* by the Germans! He may not have gotten a Purple Heart for those shins that I kicked, but I kicked him hard enough so that he should've.

"The Germans kept us marching through the afternoon. We found more and more of our division assembling. As it turns out, the Germans got their first big bag of American POWs in the opening days of what was to become known as the Battle of the Bulge. At seven thousand five hundred men, it was the largest mass surrender" of GIs in the war.

THAT SAME MORNING, Technician Fifth Grade Paul Stern and his unit of the 110th Regiment, 28th Division, was in the town of Clervaux, Luxembourg, camped out at the small railway station. They had no idea that German troops were nearby; the Second Panzer Division had raced through Belgium and Luxembourg the same way the German ground forces had in the First World War. They came up through the woods secretly, and before the GIs knew it, they were surrounded.

As he was captured, Paul was faced with a dilemma most of the other GIs didn't have. What should he do with his dog tags? His last name was Germanic sounding, but his face looked Jewish, and the tags immediately identified his religion with that *H*. For Christian GIs, keeping dog tags around their necks was second nature—if gravely sick or wounded, they might be given last rites by a pastor or priest; if killed, they might at least be buried in a grave marked by a cross. And their family would be notified of their death.

But what did the future hold for a Jewish American soldier in

Nazi hands? Paul spoke fluent German, and he knew about the Nazi brutality against Jews—from the Nuremberg Laws and Kristallnacht to the grim reports of widespread massacres and "factories of death" like the one liberated by the Soviets in eastern Poland.

A few dozen Jews were among Paul's group, and there wasn't much time for discussion about what to do. Each Jewish GI made up his own mind on the spot.

Paul didn't hesitate long; he slipped the dog tags from around his neck, smashed them with the butt of his rifle, and then quickly buried them in the muddy earth.

About 140 men were captured with Paul, mostly medics. After disarming the GIs and stripping them of their valuables, the Nazis marched them into the lobby of the nearby Clervaux Hotel. It was a high-ceilinged, upscale hotel—built in 1910—and though Paul and the other GIs were uncertain and frightened, most found the lobby a peaceful respite from the battle. They stood in their wet, muddy combat boots on the hardwood floors and wine-colored carpets, waiting to find out what their German captors had in store. Gradually the men began to sit, cross-legged on the carpet. There was little conversation.

Paul thought about his parents back in the South Bronx—he worried about how his mother would take the news that he was

captured or missing in action. He thought about his brother Jack, wondered how he was faring in France. He smiled briefly as he remembered marching in that massive parade during the liberation of Paris and then that brief reunion in Strasbourg, only weeks earlier, when it seemed as if the war was won—as if the Nazi state was about to crumble and American and British troops would be parading through the ruins of Berlin by Christmastime.

He remembered the fallout of that unscheduled visit with Jack: when he arrived back at the front in Luxembourg to rejoin his outfit, his regimental commanding officer—a major—had been enraged. The major, a medical doctor in civilian life, had stared him down. "Stern, what were you thinking? Did you get permission to go to your brother?"

Paul had said, "Yes."

But the major had barked, "Not from *me* you didn't! And I'm going to see that you're court-martialed."

"Court-martialed?" Paul couldn't understand why his major had been such an arrogant bastard, why his behavior had been so erratic and confrontational.

Shortly after the fighting broke out on December 16, Paul and the other combat medics, scrambling to help the wounded, wondered why they couldn't find enough syrettes of morphine and other necessary painkilling drugs. In the heat of the battle, they completely ran out. One night the truth emerged: the major, who controlled the regiment's entire drug supply, was caught stealing morphine. He was an addict—a junkie—and they had caught him shooting up the drug. Worse, when one of Paul's buddies was wounded in the fighting, he had to walk miles and miles without any pain medication—and ended up dying of his wounds.

Paul *hated* his major for stealing their drug supply. And he'd taken a vow that he would get even somehow.

He'd also taken a personal oath—silently, in those first days of combat—that if he was captured, he would not speak German. He didn't want to be used by the Nazis—it seemed almost a form of collaboration to relay messages to the other GIs in translated German.

They waited for about ten hours, hungry, worried about what the Nazis planned to do with them, all sitting on the floor of the hotel lobby. Suddenly, at about ten o'clock that night, a young SS man stormed through the hotel's front door, holding his service Luger in a black-gloved hand.

"Wer spricht Deutsch?" The SS man waved the handgun around as he shouted the question: *Who speaks German?*

"I was the only one who spoke German," Paul recalled. "But I just looked straight ahead. The SS man walked over to me and stuck the pistol in my chest. Of all the people in the place, with a hundred and forty Americans in the lobby, he looked towards me. I wondered who'd ratted me out."

With the Luger barrel still leveled at his chest, the Waffen-SS man led Paul into a nearby room. Paul kept asking, "What do you want? What do you want?" in English as they marched him to the adjutant of the head general of the Second Panzer Division.

"We want you to go to the commandant of the American forces in Clervaux," the adjutant said. "We have the city surrounded. You're going to tell them to surrender."

"I'm only a TF [technician fifth grade]," Paul replied in German this time. "I'm a corporal, not an officer."

"Who's your officer?"

Paul led them straight to his major, who was sitting silently among the other men in the lobby. Paul told them, in German, that this was the unit's commanding officer and that he should be the one ordering the surrender.

The SS man now trained his Luger directly at the major.

"He turned white. Yeah, he was a real bastard, my major. But I got even with him that day."

Paul wasn't about to collaborate—he wasn't going to help his German captors. No, he'd let the junkie major do their bidding.

But one question burned in Paul's mind as he waited to see

what the Nazis would do with them. "With all the GIs sitting in that lobby, how did the SS man know I spoke German?"

ALSO ON THE MORNING of December 18, the 422nd Regiment began to move from the village of Schlausenbach to a rendez-vous point in the Ardennes. While the rest of his regiment pulled out—including Roddie, Skip, and Frankie—Lester Tannenbaum was busy rigging up an army jeep—to serve as a "dummy radio station" that he hoped would trick the surrounding Germans into thinking the Americans weren't abandoning their position in Schlausenbach entirely. Lester feared it was a "casualty assign-ment," lingering behind exposed and defenseless in the city long past when every other soldier had left. He shook his head, think-ing about Bossi and Jones, the two guys who'd volunteered to stay behind with him. If Lester had had his way, they all would have been deep in the Ardennes by now, not left hung out to dry like this. It was ironic, Lester thought: Bossi and Jonesy didn't even like each other.

"Of all the guys I have to die with," Bossi told Lester, "it's gotta be Jonesy?"

But Bossi and Jones worked well together and at a fast enough clip that they were able to pull out right under the noses of the approaching German soldiers.

The 422nd had hoped to meet up with elements of its sister reg-iment, the 423rd, but when the unit arrived at the predesignated crossroads in the Ardennes, all they found was evidence of an-other hasty withdrawal. Clothing and equipment of all sorts were scattered about in the snow as far as the eye could see. To Lester, who with Bossi and Jonesy had caught up with the unit quickly, it was obvious that everyone in the 423rd had stripped down to fighting essentials and thrown away anything that might restrict movement and maneuvering. "Our boys had a field day, rummaging through the snow, replacing items they had lost," Lester recalled. "One fellow found a pass to Paris, dated for the sixteenth of December, that the recipient never had a chance to use."

Lester's unit still had its jeep, and they were able to carry more than the guys on foot; they were among the lucky ones: they had overcoats, sleeping bags, and D ration chocolate bars. The latter were getting scarce, and it had been three days since they'd had a proper meal.

As Roddie, Lester, Frankie, and Skip sat together now at the crossroads in the woods, waiting for the regimental commander, Colonel Descheneaux, to decide on their next move, the men finally had time to rest and to talk. Since 0530 on December 16 they'd been fighting against terrific odds with never so much as a whimper as to why they hadn't received any food, supplies, ammunition, or reinforcements. It was a losing fight and they all knew it. What a change in circumstances from just nine days earlier, when they'd come up on the line, a tight and carefree group, laughing, joking about the nearness of death.

Now they'd seen it—seen their friends hit. Now they knew what combat was really like. As they rested on the ground at that unmarked crossroads in the Ardennes—with nothing to do but wait—they knew fear. They could see it too, shining in each other's eyes, yet not a single man would speak of it aloud. Instead, they talked about the scruffy beards they were all growing. They talked about Camp Atterbury and furloughs to Indianapolis and the girls they'd known in England.

"The fellas know we're zipped in here tighter than hell," Skip said of that time, "yet they're all talking about our next encounter."

Maybe it was shortsightedness or bravado.

Lester preferred to think of it as *confidence*.

Maybe it was naivete.

Lester called it *morale*.

As the minutes turned into hours, the men of the convoy started to imagine getting back to rest camp, safe again behind Allied lines.

When Roddie and the other noncoms finally received word from Colonel Descheneaux, Roddie told his men, "We're going to attack—at last. We go at dawn."

At first light, the remnants of three battalions of the 422nd, consisting mainly of regimental Headquarters Company and parts of Cannon and Anti-Tank Companies, would launch an assault on Schönberg, which was now behind enemy lines, while the Ninth Armored Division launched an attack from the opposite direction. The objective was to punch through the German lines simultaneously, joining up and establishing an escape corridor for the rest of the surviving members of the 106th Division.

Because Roddie and the other men in the convoy were already in the dense woods surrounding the town, Colonel Descheneaux told them to maintain radio contact within the regiment to raise division HQ, which had withdrawn from the town to avoid being cut off. If things went bad, they were told, they'd serve as a rear guard and regimental reserve.

"To the American soldier, at least to the green combat men, once a plan is announced, it is tantamount to accomplishment," Lester wrote later. "The possibility of failure is remote and unconsidered."

They didn't move out of that crossroads until it was darker than the inside of an air-raid shelter in an English blackout. Sitting there, waiting to pull out, Lester found himself thinking about a night he'd spent in one of those deserted shelters. Only a few months earlier, but it seemed a lifetime ago. American soldiers were fascinated by those air-raid shelters, and sometimes the local girls were accommodating enough to show them around. It had been late, and Lester had been rushing back to camp. Suddenly a soft female voice had floated out of the darkness. "You will take me back to America when the war is over?" He remembered smiling at the versatility of those English air-raid shelters.

Funny, the things a soldier thinks about.

During the night, the men were ordered to take up positions in the dense woods just outside Schönberg. It was risky, moving over German occupied terrain, and they all prayed for a dark, overcast night—no moonlight suddenly bursting through the cloud cover to give them away.

Slowly the convoy moved out. Sitting in a shotgun seat, Lester marveled at how the drivers of the jeeps and trucks managed to drive without the benefit of headlights. They couldn't see the jeep in front of them. So they wouldn't ram bumpers, one GI sat in the rear seat and held up his phosphorescent watch as a guide. It was like following a trail of fireflies through a medieval forest. The men talked in whispers, continuously, stopping "only when a Jerry burp gun or rifle popped." The guns sounded so close, a few times Lester found himself reeling back, certain they were all about to be mowed down.

Somehow the convoy reached the heart of the woods without a single encounter with the Nazis. It was the middle of the night, still pitch-black, and they all piled out of the jeeps and started digging. The ground was frozen and snow-covered, so it took longer than usual to scrape out a shallow foxhole.

No sooner had Lester finished his dugout and crawled into his sleeping bag—salvaged from the detritus left behind by the 423rd a few hours earlier—than the firing began.

Tracer bullets whizzed all around them, and they knew the GIs outside the perimeter defense were firing back at approaching Nazis. It was too dark to see anything, but Lester could recognize, by the distinctive sounds, when gunfire was German rather than American.

This stopped as quickly as it had started. Lester soon learned that a small German patrol had stumbled into their perimeter, but the Germans had been driven back. There was no more sleeping that night. And Lester, Roddie, Frankie, and Skip lay there in their holes talking about what tomorrow would bring.

WITH DAWN, the distant rumble and firing increased. The morning brought Roddie and Lester the first view of their new position. Their concealment was perfect—ideal terrain, if only that patrol hadn't spotted their strength during the night.

They had to sit tight until they received word about how Colonel Descheneaux and his cobbled-together battalions were progressing. At about 0800, a courier arrived, and Roddie and Lester exchanged knowing glances: the attack on Schönberg was doomed. The Ninth Armored Division had been held back by fierce German resistance. The other battalions of the 422nd who had engaged at Schönberg had walked into a death trap, left to fight panzers with bazookas and machine guns. The supporting artillery had soon run out of ammunition. It had been a slaughter—almost every infantryman involved in the attack on Schönberg had been either killed or captured.

There they were, a roaming headquarters without an army.

Worse, an outpost of GIs had spotted a Nazi armored battalion approaching. The remaining officers of the 422nd decided to join up with the third regiment of their division, the 424th, which they hoped was still holding its position at the town of Bleialf. They were told to abandon all equipment that wasn't necessary to fighting their way out, an odd paradox for Roddie and his men because for so long they had cared for the very same precious equipment they would now blow up with grenades.

Roddie sent out their last radio message: *Destroying equipment.*

As the men went about their tasks, Lester watched in disbelief as the cooks, who "had managed to scare up some stew," dumped "that beautiful hot food in the snow."

The convoy started out of the woods, and no sooner had it hit an open stretch in the forest than the damned 88s started opening up on the road. The first shells dropped far off, but when one of the vehicles was hit with shrapnel, the men jumped out, abandoning the jeeps.

The GIs had, without planning to, chosen a good spot for a defensive stand, on the reverse slope of a small hill. Although only nine days had passed since Lester and Roddie had landed in Europe, they already knew how badly the odds were stacked against them. They couldn't fight mortars and machine guns with small arms—most of the GIs in the convoy were toting M1 carbines.

Roddie and Lester watched a group of GIs make a break for the woods across snowy terrain, only to see them cut down by machine-gun bursts just short of the tree line. Lester knew what they had been thinking because he and Roddie had discussed their chances of escaping deep into the forest and reaching the American lines. All they would need, he had reasoned, was a map of the Ardennes region. One of the jeep drivers had one but would not part with it. Lester had almost convinced him to join their breakout group when more German machine-gunning changed all their minds.

The remaining officers of their convoy—Captain Vitz and Captain Foster—ordered everyone back into the vehicles, but part of the convoy would remain behind to try to hold off the Germans. Everyone else piled into jeeps and trucks, revved the engines, and made a dash for it. All the way up the roadway's steep hill the Germans shelled them with 88s, and the men in the last remnants of the convoy threw gear and ammunition out of their vehicles, just trying to lighten their loads as their vehicles ascended. Only Colonel Descheneaux's driver refused to ditch the trailer containing the CO's clothing. Everything else, except rifles and helmets, was tossed out of the speeding jeeps—nothing mattered now except the chance to get away and fight another day.

AS THE GERMAN TROOPS in their white snowsuits drew closer, they made false promises in English, booming through their loud-speakers: "Showers, warm beds, and hotcakes!"

A single Sherman tank appeared suddenly, rumbling down the Schönberg road.

For a fleeting moment the men thought they might have armored support, but then the Nazi crewmen inside the captured tank opened fire on the Americans. All hopes of rescue were dashed.

By midafternoon, Colonel Descheneaux, commanding officer of the 422nd, could see the writing on the wall. He now had two thousand of his best men desperately ready to make a last stand. How many would the Germans take prisoner? How many would they shoot in cold blood?

Colonel Descheneaux summoned his junior officers and announced his decision: "We're still sitting like fish in a pond. I'm going to save as many men as I can, and I don't give a damn if I'm court-martialed."

The order filtered through the ranks. The men were to detonate their jeeps, to destroy all their weapons so the Germans couldn't use them. As soldiers smashed their rifles against tree trunks and tossed the last ammunition clips into a creek, one major knotted together two white handkerchiefs and set off to engage the Nazi troops.

Colonel Descheneaux was left disconsolate by the decision he'd been forced to make. He was last seen by a few GIs, sitting on the edge of a trench, weeping with frustration.

But even as most of his battalions were outgunned, outnumbered, and surrounded, some of the 422nd, including Roddie, Lester, Frankie, and Skip, desperately looked for one final way out.

A rumor had spread that some US troops had reached the town of Bleialf, where the 424th supposedly still were, and had created a narrow escape corridor. Following Descheneaux's order, they decided to try to punch through the Nazi lines with their twenty-vehicle convoy. The GIs jumped back into their trucks, jeeps, and weapons carriers and headed south. It was a furious and confused scramble. Several trucks got stuck in the mud, some overturned, and some would eventually get away.

Roddie was behind the wheel of a jeep when he spotted his buddy Sergeant Jack Sherman, knee-deep in the snow. It was very cold, and as the snow continued to fall, the Germans were getting closer.

"Jump in, Sherman!" Roddie shouted. "Stay alert. We're bringing up the rear."

Roddie hit the gas, and their jeep sped off, the last vehicle in a furiously racing convoy, trying to run through the Germans encircling them.

Sherman was a wisecracking Jewish kid from Rochester, New York. Though small and wiry, he was a tough, stand-up guy, and one of the regiment's best-trained soldiers. Roddie was glad to have him riding shotgun in that jeep.

The column of vehicles barreled through the ice and mud. In the lead jeep were three captains and two privates, including Captain Vitz, a twenty-six-year-old from Astoria, New York. The gravel road was rutted, and the tires of Roddie's jeep kept skidding.

Helter-skelter they dashed up that road, then suddenly there was a loud explosion—and Roddie saw the occupants of the lead jeep hurled into the air. Captain Vitz's jeep had hit a land mine.

Roddie and Jack screeched to a stop.

From behind the wheel, Roddie witnessed the damage. The blast had launched Vitz and the other passengers out of the jeep and into a barbed-wire fence. The jeep's driver was missing a leg, and another GI lay sprawled on the ground next to him, unconscious and hemorrhaging from wounds. Another captain and a radioman had been thrown out of the vehicle and were hurt as well.

As they tended to the wounded, the Nazi artillerymen locked in on their position, and the 88s started landing all around them. Everyone dispersed, taking cover as best they could in the trees along the slope.

From their elevated vantage point, stopped higher on the hill, Lester and Roddie could see panzers and infantry advancing. No escape maneuver was possible; all Roddie and the others could do was keep low to the ground and return fire with their .30-caliber M1 carbines.

Everywhere in the dense forest, from the snow and shadows, German troops began emerging. Germans to the left, right, and rear.

Most were wearing white uniforms and capes, camouflaging them perfectly in the misty and icy forest.

There were no lines of escape.

Roddie, Lester, Skip, Frankie, and Jack could see that they were being encircled by hundreds of Germans. They could also see other Americans being captured in the distance.

They couldn't advance; and if they tried any retreat, they'd only get closer to the German troops they could hear advancing to their rear, who were continuing to pound the surrounded men with their 88s.

Captain Edward Foster, of Knoxville, Tennessee, the highest-ranking officer remaining in the group, faced a decision: Would he let all his men be killed, or would he surrender? Some of the men felt they could make a run through the forest, but Foster wanted to save as many lives as he could. He ordered his men to destroy their rifles immediately.

Roddie loved his rifle. "I knew [it] as I knew my clothes, or anything else," he wrote, but he "couldn't bear the thought of some German using it." That image of a German using it to kill other Americans hardened his resolve. Helpless and forlorn, he destroyed the firing pin, then smashed the carbine against a thick pine tree. When the barrel finally broke free from the stock, as he later wrote, he "had to do a lot of lip biting to keep from crying." Finally—reluctantly—Roddie threw away his spare ammunition clips, hand grenades, wire cutters, and compass.

Meanwhile, all the Jewish infantrymen were faced, as Paul Stern had been, with another life-or-death dilemma: Should they rip off their dog tags and bury them in the deep snow? Most knew they could not risk capture with that *H* for Hebrew identifying them as Jews. Some Jewish GIs swapped dog tags with those around the necks of dead comrades.

"One of the things we were told was if we were captured, we should destroy our dog tags," Lester recalled. "When Captain Foster ordered us to surrender, I destroyed everything. I destroyed my dog tags even before I destroyed my rifle."

Skip Friedman, on the other hand, kept his tags around his neck. He later explained, "When we were captured, many of the Jewish POWs threw away their tags. I was really a little pessimistic about what was going to happen to us—so I decided to hold on to my dog tags so [my body] could be identified."

As each group of GIs surrendered, the Germans searched the prisoners, taking watches, rings, any other jewelry they could find. They also stripped the GIs of their overshoes and winter coats.

When there was a burst of loud shouting among the Nazis, Roddie turned to see a German sergeant arguing with a lieutenant.

Nearby, Technician Fourth Grade Hank Freedman, raised in an Orthodox Jewish family in Boston, was listening carefully. Freedman was short and wiry and had grown up with Yiddish as his first language. Hank managed to decipher the Nazis' argument.

"Hank, what're they saying?" Roddie whispered.

Hank's eyes darted between Roddie and the Germans, and he responded with a grim expression.

"They're debating what the hell to do with us," Hank said.

The sergeant was agitated, eager to get going. Rather than take the Americans as prisoners, he wanted to shoot them all. He kept saying they were on a timetable. "We can't waste this time with these vermin. Shoot them and let's go!"

Roddie watched the Nazi lieutenant's face closely—he knew that his life, the lives of all in his company, hung in the balance and came down to this one young Wehrmacht lieutenant's judgment.

For his part, as a man of devout Christian faith, Roddie had every intention of surviving. Even as the two Germans continued to argue, he took all the US cash he had hidden in his pocket, folded it into a tight square, tore a slit under his shoulder patch, and secreted it away. He knew that greenbacks might come in handy in an escape from captivity.

IN THOSE SHORT DAYS of December, it was already dark by 6 p.m. It was a bitter-cold, snowy, windy night.

The Germans ordered their prisoners to lie down. "Sleep in the snow!" came the order.

Ultimately the men were forced to lie in a muddy, frozen churchyard, uncovered, in a nearby town, where they formed themselves into pyramids of shivering bodies.

"Come on," Roddie said to Jack Sherman. "Let's try to get some sleep."

Huddled together, bodies pressed tight, they tried to share whatever body heat they had left. All night, captured GIs joined Roddie and his boys.

The courtyard of the church, which had been built in 1380, was surrounded by guards and growling German shepherds trained to kill. The Germans had their Maschinengewehr 42s—machine guns known to the US troops as "Hitler's buzz saws"—trained on the captured men throughout the night.

Skip Friedman felt enormous dread when he had entered the churchyard. "When you're in brutal hands like the Germans, you're scared as hell," he said later. He too joined that mass of infantrymen, huddling together, trying—without success—to sleep.

Besides freezing to death, the worry of every GI was that the Germans might suddenly machine-gun them. The atrocity at Malmedy had been committed not far from this churchyard.

That fear of a sudden burst of machine guns never left their minds throughout the night.

FIFTEEN

A T DAWN ON December 20, the Nazis shouted for the POWs to get up.

Only now did Roddie and Jack Sherman get a full understanding—in the stark winter daylight—of the carnage.

And then they began a long and terrible march.

Roddie marched stride for stride with Jack. Along the route they saw the corpses of guys from their own outfit—fellow 106th Division infantrymen—faces twisted and frozen in the moment of agonized death. Roddie and Jack glanced at each other, flinching. They recognized several of the dead. Those motionless, frozen faces had been friends. Guys with whom they'd trained, guys they'd heard laughing, joking, talking about their families, only months before at Camp Atterbury in Indiana and at Guiting Grange in England. Now they weren't moving—just frozen corpses.

All the American trucks and jeeps had been burned. The Sherman tanks had been burned. That was all that remained of the American lines.

Every mile or so Roddie heard a blast from one of their captors' rifles. Resting, straggling, or trying to escape meant summary execution.

The Nazis forced the prisoners to march with no regard for their well-being. John Morse of the 423rd Regiment later said, "We were lined up, stripped of all but clothing, and marched off. What a sad, disappointed, sorry lot we were. Helping the wounded along, we merged with other new POWs to form a ragged phalanx of shuffling zombies."

Stopping for any reason meant risking their lives. They marched without food or water. Still, Roddie and a few of the men managed to grab some sugar beets they found along the side of the road, quickly taking a bite and handing the leftovers to the shivering GI next to them. Some surreptitiously crouched down to cup their hands in dirty puddles and sip some water. If the Nazi guards saw any man sipping from those puddles of blackened, oily ice water, they rifle-butted the GI or threatened to shoot him.

Besides trench foot, many of the troops were suffering from numbness and felt sharp pains in their extremities. Roddie was developing frostbite on his toes. But slowing down wasn't an option.

The wounded, Morse recalled, "suffered the most. Our medics, and others, helped them along because to fall out could be, and was in some cases, fatal. The shots we heard from the rear testified to that."

Skip put it succinctly: "If you didn't march, you didn't last."

In addition to the cruelty of their captors, the POWs were vulnerable to civilian attacks and even lynch mobs. Several of the POWs marching in columns recalled seeing downed Allied airmen who had been lynched from lampposts by enraged Germans.

In Koblenz, a civilian in a business suit ran up and hit one of the

GIs in the head with his briefcase. The German guard nearby said the man was upset over the recent Allied bombings.

Up hills and through pastures covered in snow, slush, and ice, they marched. Panzers, artillery, thousands of Wehrmacht and Waffen-SS men hurried past them to the front to take part in the desperate Nazi push toward Antwerp.

"As our march into Germany continued, we went through Prüm," Morse said. "Along the way German troops moving up to battle would stop some GIs to search for watches and other souvenirs. Our combat boots attracted a lot of attention. A German would place his boot next to a GI's and, if the size looked right, demand a swap."

On the narrow roads, the columns of POWs often had to jump out of the way as advancing German tanks and heavy artillery barreled past them, heading toward the front. "A couple of fat German officers almost ran us down driving through the column in one of our jeeps that they had apparently captured early in the battle," recalled Richard Peterson, a sergeant with the 423rd Regiment. "The jeep still had the division HQ designation on the bumper."

Not every German guard was equally brutal. Some young enlisted German soldiers, in fact, seemed resigned to the desperation of their own fate. "There were distinct differences in the German Army," Skip recalled. "The Wehrmacht were ordinary guys who were drafted, hauled in, and the SS was always in back of the Wehrmacht. The Wehrmacht either kept advancing or they got shot by their own people. That's the way the German Army worked. And as we were marching this regular German soldier comes up to me—he spoke good English; it turns out he was a guy that had lived in New Jersey—and he said, 'You guys are lucky. Your war is over. We've got to go on until we get wiped out.' That's what he reported to me. And God, he was right."

The German Army troops who had guarded the POWs on the road to Prüm were replaced by older men of the Volkssturm, a people's militia. Some of them were wearing pilfered American army boots and overshoes.

Like every other POW, Roddie was uncertain. He was freezing, miserable, and starving. He didn't know how long they would march—or where—though gradually he and the other men started to hear rumors.

"We heard we were being marched to Gerolstein," Skip said. "That's the first time I ever was introduced to a place named Gerolstein."

The men all quickly learned that Gerolstein was a major railway hub, which could only mean the POWs were being taken farther away from the front, deeper into the Nazi Reich.

"We were marched 31 miles without food and water," Roddie later wrote in his diary. "We were herded into a lot and slept or lay in the mud until morning. We were then given two bags of hardtack, the most distasteful crackers I have ever tasted, also some cheese. This was the 21st of December. Our first food."

For the first time men of the 106th and 28th Divisions were mixed together. Another dreadful night sleeping in an open encirclement, surrounded by barbed wire, in the freezing air. Huddling together and sleeping on top of each other for warmth.

As they lay in the mud at Gerolstein, Roddie and every other GI slowly came to the realization that there would be no relief, no rescue, from their fate. "We would remain as captives of our enemy," Richard Peterson said. "No cavalry would come charging over the hill. Life would go on until God or the Germans decided to end it."

PART IV

The fear of God makes a hero,
the fear of man makes a coward.

—REV. A. C. DIXON

SIXTEEN

A T DAWN ON December 21 the order came for the men to
form ranks and march into the Gerolstein rail yard. Disori-
ented, as dogs barked viciously at him, Roddie was herded with
other men into boxcars, which he knew were destined for Germany.

It was freezing cold and windy, and the snow was thick on the
ground. Lester was lucky enough to have kept his overcoat—not
many of the men had. Like most of the GIs, Lester had lots of water sloshing around in his boots, and he realized that his feet were
at risk of freezing. Many men were without helmets; Lester had
thrown away the metal part of his and wore just the plastic helmet
liner. The protective metal of the helmet had been too heavy to
wear all the time.

The boxcars were somewhat smaller than those of US freight
trains. Most were filled with hay and covered with animal excrement.

"We were packed in like sardines—no food or water," Skip said.

Boxcars that could hold four horses, or perhaps thirty men, were crammed with seventy-five POWs or more. Many of the men were wounded, moaning, suffering from trench foot and frostbite. Some resembled sacks of coals more than men, slumped over in starvation and exhaustion.

"The boxcars turned out to be the same type of boxcars that were taking the Jews to concentration camps," Sonny Fox said.

There were only four tiny openings in each boxcar, two on each side at the top. That was the only source of daylight, and each of the openings was covered in barbed wire, "rendering thoughts of escape hopeless." Though the GIs didn't realize it yet, no "POW" signs were painted on the boxcars—as required under the Geneva Convention of 1929—so any Allied pilots could easily assume the trains were carrying artillery, materiel, and German troops, not unarmed prisoners.

The sanitary conditions inside were appalling. The stench of urine and feces was overpowering. Soldiers relieved their bladders against the wooden walls, or sometimes right where they stood. Many of the GIs were suffering from dysentery. The men used steel helmets to empty their bowels, then passed the helmet along until it reached the GI closest to one of the thin openings, who then tried his best to dump the contents out into the freezing air through the barbed wire.

Roddie tried to maintain discipline, but the cramped and filthy conditions made it next to impossible. The men were imprisoned in a rolling hell, without any sense of a final destination or when they might eat or have a drop of water. Desperation filled the eyes of some of the men.

"We couldn't sit down, couldn't lie down; we had to stand there," Lester said. But for some of the prisoners, "the mood was surprisingly one of laughter. We kept telling stories that made us laugh. It's the thing that kept us going even though we were starving. We'd be talking to each other about our experiences—not so much about combat, but what life was like back home. That's what we were living for."

The trains moved slowly, sometimes pulling to a standstill off the main tracks for minutes or hours. Every time a German military train came along, the POWs found they'd get shunted to a siding track.

Once, while the train filled with men of the 106th and 28th Divisions waited, a train carrying panzers and their crews stopped on the next track. Richard Peterson recalled being able to see the faces of the German enemy clearly—he and the other Americans made sure not to show the Nazi troops even a hint of their desperation. "We traded insults in English and German without understanding anything but our mutual hatred," he recalled. "They looked so damned cocky and healthy. . . . I began to realize the benefit of the discipline under which we had trained and lived. The result was men who were not only properly dressed but reflected dignity in their carriage and expression. Only when we yelled out that George Patton was looking for them were there any signs of acknowledgment. I prayed they would meet the general's troops at some point and he would destroy them as they had destroyed so many of us."

In truth, most of the infantrymen held out little hope of rescue—from General Patton's armored division or anyone else. Their entire existence became staving off the pangs of hunger and thirst, and withstanding the stench and "claustrophobic hell," in the words of Sergeant John Kline of the 423rd. "Some men leaned against the sides of the car—hollow eyed, faces blank. Others sat, weak from wounds being tended by their friends. A few sagged glassy eyed from fatigue or to avoid their desolate surroundings."

The gloomy overcast skies offered little light, making day virtually indistinguishable from night.

ONE WEEK AFTER the Ardennes counteroffensive had begun, on the morning of the twenty-third, the bad weather that had served the German Army so well during the initial days of the battle gave way to bright blue skies—and now the British and American airmen could fly reconnaissance and bombing missions with virtually no risk from Luftwaffe fighters or German anti-aircraft. The Royal Air Force launched a massive nighttime raid with some three thousand bombers to destroy industrial and military targets in Germany.

The Allies also could now send in relief to the troops still fighting in Belgium, desperately defending Bastogne.

Roddie was most likely thinking of Christmases back in Tennessee, walking with friends and family to the Methodist church just a few blocks away from home. The church, a small redbrick building that to Roddie's young eyes—as he recalled in his diary—resembled a castle, would be as dark and lifeless as the boxcar. But it slowly would come to life, glowing with a blaze of candles and the joyous sounds of carols. Young Roddie loved singing in unison with his father and aunt, brothers and cousins.

Silent night. Holy night.

The image was suddenly shattered by the buzz of aircraft coming in low. RAF planes—normally a sound the infantrymen would have greeted with whoops of joy.

But now the men locked in the boxcars started to shout in terror.

They heard a hissing sound, and a flash of lights shot through the barbed-wire slits in the boxcars. Flares dropped from the RAF planes lit up the rail yard as if it were daytime, guiding the way for the heavy bombers to unleash their loads.

Fifty-two RAF Mosquito fighter-bombers, on a mission of strafing, were flying in formation to level the Limburg railway station.
Adjacent to the boxcars filled with POWs were railcars laden with
heavy artillery destined for the Nazi march to Antwerp.

Panic started with the first explosion and reverberated, seismi-
cally: tremors building to an earthquake.

Another explosion, closer—this time a direct strike on one of
the forward cars.

The men heard the nearby screams of their fellow GIs, just as
trapped as they were. Boxcars destroyed by the RAF bombs de-
railed, bursting into flames.

Some of the POWs had busted out of the damaged cars and
were killed while trying to escape the carnage. "The earth shook
with explosions and the unforgettable whistling of more destruction to come," John Morse recalled. "Nearby a freight car was hit.
Through the cracks in the wall we could see prisoners run from
the wreckage only to be shot down by the guards."

It was a moment of devastating fear and powerlessness.

"The British were bombing the rail yard: you could hear the
bombs whispering down," Sonny Fox later said. "When you
can't run, you can't hide. In combat at least you can move, dig
in, seek cover or shoot back. To be sitting there, listening to the
bombs coming down, with no alternative but to wait and hope
each one missed—seemed to make the word 'hopeless' totally
inadequate."

In Roddie's boxcar that hopelessness turned into confusion and
terror—then violence. Men clawed at one another and tried to
rip through the wood with their bare hands, desperate to escape.
Some of the prisoners tried to break through the barbed-wire-
covered vents with their hands, bloodying their fists.

"From the other side of the boxcar I heard a voice with a Southern drawl rise up above the noise and chaos," Hank Freedman later said. "It was Roddie. This man comes right out and says, 'Boys.' The minute he said, 'Boys,' we listened. Roddie had a commanding voice. Very much under control. It was a voice you didn't mind listening to. You wanted to hear what this person had to say.

"And Roddie said very calmly, 'Boys, if you have ever prayed to God, you need to pray now and ask him for salvation. Have faith! God will save us. Pray, boys, pray!' And we got quiet and we prayed. What he said set the tone."

"God will save us. Pray, boys, pray!"

A calm came over the car, Hank remembered. The men stopped struggling and flailing.

Have faith!

Hank was astonished at the transformation when Roddie bowed his head and prayed aloud. Outside, the bombing continued: wounded Allied prisoners were screaming, running for cover that they could not find.

"Pray, boys, pray!"

Bombs falling outside destroyed more trains, killing more unarmed Americans.

Inside Roddie's boxcar, the fighting had stopped—the terrified men stayed disciplined, followed his orders, and silently prayed.

Roddie had never prayed as fervently as he did throughout those terrible moments of the RAF bombing. It seemed as if each bomb was coming directly at their boxcar. Roddie's boys astonished him with their bravery. He was proud: they obeyed his commands even in the face of death.

And their prayers were answered. They gradually heard the bombing subside. Men were screaming nearby—the wounded and the dying—there were periodic sounds of rifles firing, but the horrible concussions and tremors of the bombing had ceased.

In just a few weeks on European soil, Roddie had survived an artillery barrage and a mortar barrage, been shot at and brutally treated by German troops, and now bombed by Allied planes.

None of it "was gravy," he thought, standing and shivering in that dark boxcar, but this bombing was by far the worst.

SEVENTEEN

A FTER THE BOMBING, the men remained locked in the boxcars—traumatized, relieved they were still alive. But then the pangs of hunger became overwhelming. The train showed no sign of moving, and the men continued to hear the groaning of the injured. They were startled when they also heard the door of the boxcar being unlatched.

A German guard opened the door, threw a loaf of bread inside, and said, "Merry Christmas," in heavily accented English.

Roddie and the other guys were a little astounded. Sixty starving men and one loaf of bread between them. All eyes turned to Roddie—he was the ranking NCO.

It was their first sight of the "bread" with which they would soon become all too familiar. Smaller than an American loaf, it was a black bread made with a copious amount of sawdust as filler.

After much discussion, and a lot of unsolicited advice on how to divide up that meager ration, Roddie took the loaf and placed it in the palms of the soldier next to him. Using his dirty fingers like a makeshift saw, he managed to tear it into sixty pieces with the precision of a diamond cutter. Each piece was tiny—no larger than the square pat of butter you'd get in an American restaurant. Roddie felt like the fabled Dutch kid who stuck his finger in a dyke.

Ernie Kinoy, a private in the 422nd from New York City, told Roddie he was astounded to see him divide such a small loaf of bread evenly into sixty portions.

After the morsels of bread had been distributed, Roddie started singing "Silent Night." A second GI soon joined him. Then a third.

The POWs heard angry words shouted in German and rifle butts banging on the boxcar door. The men stopped singing and crowded shoulder to shoulder in the darkness. Complete silence.

"To hell with them—it's Christmas Eve!" one of the GIs shouted, and he began singing with as much gusto as he could muster. In no time, the entire boxcar was singing. Roddie's voice rose strong in the chorus.

The Christmas carol was, in its way, John Morse recalled, "a moment of rebellion. At the end of our song we fell silent again, having demonstrated our independence."

A few moments later they heard the sound of several soldiers singing in German the same melody of "Silent Night."

Stille Nacht! Heilige Nacht!
Alles schläft. Einsam wacht.

Some GIs sang all the Christmas carols they knew and were echoed by their captors with the German renditions. "The words were different but the songs—and maybe the emotions—were the same. For a few brief hours there was peace on earth among us there in a ruined freight yard," Morse said.

For Lester Tannenbaum that night's caroling had a far darker tone. That Christmas Eve he heard Nazi troops nearby singing:

O Tannenbaum,

O Tannenbaum,

Wie treu sind deine Blätter!

"O Tannenbaum"—that classic Christmas carol—now had a "terrible association," he said. Lester vowed that if he survived this boxcar hell, if he survived *whatever* lay ahead and returned to New York, he would legally change his surname to "Tanner." That beautiful, poetic-sounding word *tannenbaum*—"fir tree" in German—was marred forever. Lester never again wanted his name linked with that German Christmas carol the Nazis were singing just outside the boxcar.

ON CHRISTMAS DAY, after five days locked in those boxcars, the trains crammed with POWs finally stopped in a small town. The doors flew open, and armed Wehrmacht troops shouted, *"Raus! Schnell!"*

Roddie, Lester, Frankie, Hank, Skip, and the other men fell out of the railcars, their limbs freezing, barely able to stand. They were filthy; none had washed or shaved since the battle began on December 16. Their uniforms were fouled from the dirt of the battlefield and excrement. Some no longer looked like soldiers but like the unshaven, bedraggled European refugees they'd seen in newsreels.

The train had stopped short of the depot and perhaps, one POW said, in a "deliberate attempt at dehumanization" by the Germans, the men were forced to march into the town.

In a long, snaking line, they marched past the three-story red-stone station house. A clocktower rose from the middle of its roof peak. Over the main entrance, a large sign with black letters identified the station as Bad Orb. Few of the men knew that *bad* meant "spa" in German, but they knew what it meant in English.

Bad Orb is a picturesque resort town in the state of Hesse. The geology has dictated its centuries-long popularity as a destination: several sulfur hot springs said to have curative properties attract visitors to the various Bad Orb baths. Roddie and the men didn't yet know that since 1939 Bad Orb had also been the site of Nazi Germany's largest prisoner-of-war camp, located on the nearby hill known as Wegscheideküppel.

"Going through the lovely little town of Bad Orb," Skip recalled, "kids were singing Christmas carols, lights were out on display, and . . . we were grim as hell."

They marched along Bad Orb's cobbled streets, past gabled homes, inns, and churches, and the road twisted uphill past a German Army medical facility.

"On our right was a huge German amputee center," Skip said. "Dreadful. Hundreds of amputees from the German Army on crutches walking around. I saw severed arms and legs stacked on the side of the building. One of the most grotesque scenes I've ever seen. There were literally hundreds of people walking around without arms and legs. It was gruesome."

The men could barely function with lack of sleep and food, but now the rigorous physical training Roddie and other instructors had put them through in the woods of South Carolina, the hills of Tennessee, and the prairie of Indiana paid off. They marched along solely on what one GI called "built-in reserves of self-discipline."

A downcast Richard Peterson trudged along too, exhausted, raising his eyes "to find a crystalline scene surrounding us. Even in the low-hanging sunlight of winter, the sky had an almost painful blue intensity. All around was the clean green of pine trees sharply outlined against the blue sky and the new white snow. Even the ground looked soft with its winter blanket. The beauty of the vista pierced through my misery. 'Look around,' I said to the man next to me. 'If I weren't so cold, so hungry and so scared, this would be one of the most beautiful places I have ever seen.'"

To Roddie the beauty must have been doubly painful; the landscape looked just like his Smoky Mountains and the steep, winding road to Cades Cove. For an instant, he escaped the hell he was in. It seemed like he was home.

Roddie and the POWs made an arduous five-mile hike from the freight trains, up a steep, zigzagging, icy mountain path. It would have been a difficult march for well-fed healthy men. Along the way, Paul Stern and others cupped their hands and filled their mouths with snow—they hadn't had any water now in days.

At last they saw their destination: Stammlager IXB—or Stalag IXB. Originally a World War I army training camp, it was turned into a children's summer colony after the war. During the 1930s it secretly became a training facility once more, for glider pilots, a ruse that saw Hermann Goering's Luftwaffe emerge before World War II. In the early 1940s, thousands of French prisoners of war filled the barracks. It was a massive camp, known for having some of the most unsanitary, disease-infested conditions of any Nazi POW camp. By the time Roddie and the men arrived, Stalag IXB housed prisoners from at least eight countries—France, Italy, Great Britain, Belgium, Serbia, Slovakia, the Soviet Union, and the United States—with a peak population in late 1944 of more than twenty-five thousand POWs.

The men plodded past the commandant's three-story mansion, then the kitchen, manned by haggard-looking Soviet POWs. They saw that most of the wooden barracks had broken windows.

Roddie and the other GIs were shocked when they saw the camp latrine, which was simply an open pit, about six-feet square, uncovered, surrounded by low poles on short supports the men were expected to balance on while using the facility. It was another reminder that they were moving, inexorably, away from being seen as proud fighting men toward what their captors viewed as animals, or "sub-humans." "The Nazis' goal," Skip later said, "was our humiliation."

They were formed into ranks for the now-standard *funf-mann*—or five-man count—so the Germans could begin an official documentation of their imprisonment. Roddie and most of the others understood that this was necessary—it was the only way their families back in the States could be notified that they had not been killed in action but rather were POWs in Germany. Each man learned the German name for prisoner of war: *Kriegs-gefangene,* or *Kriegie,* in the jargon of some of the older prisoners. One by one, they were interrogated.

EIGHTEEN

I N GERMAN CAPTIVITY, prisoners were segregated not only by branch of service but by rank. There were dozens of *Mannschaftslager* ("Stalags") for private soldiers and NCOs, and *Offizierslager* ("Oflags") for commissioned officers. Life as a POW for an officer was far different from life for anyone under the rank of lieutenant. Sergeants, corporals, and privates were segregated and sent to separate camps. Noncoms were sent to camps designated with an *A*, while privates and privates first class were sent to camps designated with a *B*. Furthermore, the Geneva Convention allowed for different treatment of noncoms and enlisted men: officers and noncommissioned officers were not obliged to work as prisoners, but the Nazis could put privates to work.

Roddie and many of the others had committed to memory the protections they were supposed to be granted as POWs. The War Department had issued all men a pamphlet titled *If You Should Be Captured, These Are Your Rights*. "When you are questioned," the pamphlet instructed, "by no matter what enemy authority, you

must give only your name, rank, and serial number. Beyond that, there is no information [that] the enemy can legally
force from you. Do not discuss military
matters of any sort with anyone."

Upon arrival, the new POWs were
forced to fill out two information cards,
one for the International Red Cross and
another for German military records. The German captors, who
issued all the men *Kriegsgefangene* ID numbers with dog tags, also
asked the men to fill out questionnaires with personal information, in clear violation of the Geneva Convention. "One of the first
things they did was to ask us our whole family history, what our
family name was, where we came from, the names and address of
our relatives and things of that nature," Hank Freedman recalled.
"Of course, we didn't tell them. We said, 'Here is my name, this is
my rank, and this is my serial number, and that's all you're going
to get.'"

Hank, Roddie, and Frankie were among the group of twenty
GIs who flat-out refused to fill out these questionnaires. As a consequence, the Germans ordered them to strip naked to the waist,
then forced the men to stand outside in the brutal cold for more
than an hour and a half. The men shivered, rubbed their hands
on their arms and shoulders, stamped their feet to try to delay the
onset of frostbite—and tried not to let the Nazi guards see any
signs of their discomfort.

Hank recalled, "An American officer who happened to be there,
I guess, took pity on us, standing there half naked, and he said,
'Look, guys, it's stupid, and it's ridiculous for you to persist in not
giving them the information.'"

Roddie didn't agree with the officer's decision, but he also didn't
want to see the other GIs suffer needlessly.

Beyond details of family life and addresses, the Germans demanded one other piece of information for their files: *religion*.

This, of course, posed a terrifying dilemma for the Jewish GIs.

Lester, Skip, Paul, Sonny, and Hank weighed the decision carefully. Should they proudly say, "I am a Jew"? Should they trust they would still be treated as United States soldiers, even in Nazi captivity?

Those who had destroyed their dog tags with the telltale *H*, or swapped dog tags with those of dead comrades, had the option to deceive the Germans. Lester told them he was Presbyterian. "Very few of the Jewish prisoners identified themselves as Jews," he explained. "We were too smart for that."

Private Ernie Kinoy agreed. "I was politically aware enough that I was not about to register with the Germans as Jewish."

Yet some, like Skip Friedman, had held on to their dog tags—and had an awful decision to make.

Sonny Fox recalled getting to the front of a line and encountering an American POW designated by the Germans to assist with the registration process.

"Name?"

"Irwin Fox."

"Rank?"

"Sergeant."

"Serial number?"

"42022375."

Sonny was a bit taken aback when he was asked to state his division, because the Germans already knew they had attacked the 28th Division, and the red keystone patch on the shoulder of his combat jacket gave it away.

"Company E, 110th Regiment, 28th Division," Sonny said.

"Father's name?"

"I'm only supposed to give name, rank, and serial number."

"Listen, kid, if you don't answer these questions, they're going to make you stand in the snow until you do."

Simply wanting to get out of the cold, and get some food, Sonny figured that the Nazis were "not going to win the war" if he gave them his father's name.

"Julius Fox," Sonny said.

"Occupation?"

"Textile converter."

"Religion?"

At first Sonny refused to give an answer.

"Religion?" The clerk was waiting impatiently.

"Jewish."

The clerk looked up from the form, stared at Sonny for a moment, and said, "Protestant."

At first, Sonny thought he had misheard him. "Jewish," he repeated.

Then, finally, Sonny understood what was going on. He was a Jew in Germany. Why place himself in peril? The clerk was only trying to protect him.

But Sonny wouldn't change his answer. He had already surrendered, thrown down his rifle. Now he was supposed to renounce his faith, surrender morally after giving up physically? He was cold. He was hungry. He was exhausted, too tired and too beat-up to be devious.

"Jewish," he repeated.

This time the clerk didn't even look up. "Protestant!" he said, waving Sonny away. The man wasn't about to waste any more time on some fool—a putz, Sonny thought—who didn't appreciate what he was trying to do for him. He told Sonny to move on, which Sonny did, gladly, quietly relieved he wouldn't have to stand outside in the snow like some of the other GIs.

Inside the camp, the barracks were badly overcrowded. Most

housed up to one hundred men. Broken windows allowed in the bitter cold. The bunks had three tiers; some had no bedding, merely wooden slats. The few available mattresses were really just "filthy pads filled with dried grass and ferns." There were no blankets, and the tiny stove in each barrack was devoid of fuel, providing no heat. The GIs had only their field jackets, no overcoats, gloves, or hats. Dirt was ground into the floors. Other than the bunks, there was no furniture. Despite these conditions, and the freezing temperature, the men collapsed out of sheer exhaustion wherever they could find a space.

At least they could huddle together for warmth. Some men sat on a buddy's frozen toes to try to stave off frostbite.

According to Skip, it was part of the Germans' overall strategy to "dehumanize" the POWs "and turn them into animals," but the men did their best to try to maintain a sense of morality.

Staff Sergeant Earl Verham, a medic in the 28th Division, recalled what the chaplain at Bad Orb had told him: "It isn't what happens to a man that matters. It's how a man reacts when something happens to him that matters." To Verham, the men in the prison either became animals or proved to be men. "Stealing bread was taking a man's blood. Yet it happened." For punishment of theft, the POWs organized makeshift courts. "We pronounced sentence and kept a record, which was turned over to our American officials on liberation. As an example, if a slice of bread was stolen, the sentence would be one week of latrine duty and a recommendation for reduction of rank when liberated."

The smallest things could help keep a man sane. For Frankie it was washing out his undershirt, socks, and pants to feel at least somewhat clean. Sonny took the advice of an older POW in the barracks, a man of probably twenty-eight years of age. "'I want you to do two things,' he told me. 'I want you to shave, and I want you to wash out your pair of socks.' It seemed like such a silly thing to ask. Who the hell cared? It wasn't as if the USO was coming to the stalag to entertain us. Every day in camp was the same. We were all in the same unshaven, neglected state. But I respected this man and decided to do it."

There was just a single razor blade for 240 men. There was only a cold-water tap in the latrine. "To shave meant that I had to stand there, in the freezing weather, in the cold outhouse, shaving with a dull razor. But I had committed myself to doing it, so I stood there on a cold January day in the outhouse and shaved. It took about twenty minutes, and every time I cut a follicle, it brought tears to my eyes. By the time I finished shaving, my hands were frozen, my fingers about to fall off. But I did it."

Sonny said he decided "not to wallow. I had taken back control of my life. Yes, I was still a prisoner, but in some very important place inside of me the barbed wire had come down. I had learned that even under the most hopeless of circumstances, one does not have to be helpless. I learned another valuable lesson through this experience: man cannot exist without a social structure. Until I entered military service, the social structure I grew up with was the civilian one. When I was inducted into the army, I entered a social structure that was totally military based. Rank dictated who we obeyed and to whom we deferred. When I became a sergeant, I automatically assumed power over eleven other soldiers. They were not to question my competence or my abilities. I had three stripes and that was it. And when we entered the stalag, we were supposed to continue to treat our community as American soldiers."

However, for the men's first few weeks in the camp, Sonny said there was no recognizable social order. "Gradually, a new structure valid for this peculiar situation emerged. Language facility helped. The ability to give a haircut was valuable."

As was tailoring. "Surreptitious looks at comrades emphasized the sad knowledge that we were slowly deteriorating physically,"

Richard Peterson recalled. "Skin was gray, hair dull, and eyes expressionless. Clothes hung loosely on our bodies. While our hair did not grow very much, every head was filthy. The gauntness in young faces was scary. A full-length mirror was not needed. One could see himself in the condition of his comrade. . . . As men lost weight, their pants got too big. The task of tailoring was open, so I took it. A needle came from somewhere. Thread was reused from the pant seams. It is fascinating what one can use when there is no other choice. The job offered something useful to do with my time and allowed me to be alone with my thoughts."

The POWs quickly got used to the mind-numbing and grim routine of camp life. But they could do nothing about the starvation rations doled out by the Germans.

Frankie wrote in his diary: "Our daily diet consisted of breakfast—ersatz tea with no sugar whatsoever, dinner ¾ of litre of inedible soup. Supper ⅙ of a loaf of bread and approximately 1½ square[s] of potato margarine."

At 0600 sharp the barracks lights came on and shouts of *"Raus! Raus!"* from the German guards followed. Once the men were all accounted for, each group dispatched a two-man team to the canteen to fetch the morning "ersatz tea," which was actually just dark roasted grains of some sort that had no flavor. "The only good thing one can say on behalf of this brew is that it was warm," recalled one POW.

There was nothing provided in the way of bowls, cups, or spoons. The men had only their helmets in which to receive the "tea"—more often described as "coffee." Other than its color, the brew bore no resemblance to actual coffee. Because it was served hot, many men used it for shaving.

At noon, a different two-man team would fetch from the kitchen a concoction called "grass soup," which was served in a tub. The men ladled it into their steel helmets, just as they would the coffee. Some, like Lester, converted their plastic helmet liners into all-purpose bowls. Others drank straight out of their steel helmets.

Both the soup and the tea/coffee had nearly no caloric value; the main meal was a portion of black bread—a miserably small portion for each man but essentially the high point of the day. To Ernie Kinoy, the tiny sliver of "sawdust bread" looked more like "a paving block than anything edible." It was of negligible nutritional value. It's estimated that most of the POWs in Stalag IXB received between 500 and 600 calories per day.

Lester would write later that his most vivid image of camp life was the way the men divided that black brick-like loaf between six men. Every group developed its own system.

"That small loaf of what the Germans called bread, our meager daily ration, was our only solid food," he said. "It tasted mostly of sawdust. Six men shared one loaf and we took turns in cutting it, with that person being the last to choose from the six pieces. No arguments [occurred] but [there were] different ways of consuming that piece, quickly or saving some for later. Some men liked to toast their piece over a flame, and then there were men who would trade parts for a cigarette or two. We talked about food all the time, not the usual banter about women. The hunger, which was ever-present, drowned out the sex drive. Everyone lost lots of weight—I went from 180 pounds at 6 feet to 120 pounds in 100 days."

Other groups drew lots, or cut a pack of playing cards, like John Morse, whose group had to divide their slice eight ways, rather than six: "Some way had to be devised to be fair," Morse wrote. "I made a knife from half of a door hinge I prized from the door of a small room at the end of the barracks. Our walls were of stone so I sharpened my 'knife' there. We found a deck of cards I used to 'cut for choice.' High card chose first and so on down the line. I would carefully measure eight slices, cut the bread and the fun would begin.

High card man would step up and eye the pieces to see which might be a tad larger. And so it went. At least nobody felt cheated."

Sonny recalled, "The bread started out being shared by five men, then seven and eventually was reduced to being shared by ten prisoners. We developed a routine to ensure fairness in the selection of the pieces. Each evening became a kind of vespers ceremony. We organized ourselves into communities of ten and each night a different member of this group would slice the bread. He would cut the slices as equally as possible, and then lay each slice facedown to hide any air holes which might lessen the quantity of bread one would be getting. The next step was to draw from a deck of cards. This would dictate the order in which we selected our slice. Obviously, the preferred position was first, so if any slice looked larger than the others, one could lay claim to it. The ceremony of the bread took longer than the actual consuming of the slice."

The process of cutting, choosing, and consuming that tiny portion of sawdust bread was a highlight of most POWs' days.

Frankie described the ritual in his diary: "Louie, myself, Paul (a new and good friend) and Roddie had coffee every night with our bread. It's really something to watch us when we get our bread. We cut it up into small paper-thin squares and toast it on top of a small stove we have in the orderly room. We look forward every day to our night together when we toast our meager supper and chat about home, our loved ones and food."

It was inevitable, of course, that some fighting occurred, but with the POWs having so little strength, the few fistfights that broke out rarely resulted in blows being landed. Most, in fact, ended with little more than hollering, feeble haymakers followed by half-hearted handshakes.

"A friend and I got into a violent shouting match one day over some real or imagined insult," Richard Peterson wrote. "We decided to settle it outside with fisticuffs. After he swung at me once and I retaliated in kind, we realized that further fighting was use-less. The effort to swing was too great, even if we landed a blow it would be ineffective. We fell into each other's arms laughing at our ridiculous behavior."

To keep the men occupied throughout their days of captivity, other sergeants in the self-appointed "committee" were assigned the task of organizing religious groups and educational programs. Anything to take the men's minds off the ever-present hunger. Frankie recalled finding solace in daily prayer and Roman Catho-lic services with other POWs. He wrote in his diary, "I took some time each day to read Roddie's Bible and recite my rosaries." In an-other entry, he wrote, "What a Sunday this has been. This morn-ing I went to Mass and received Holy Communion. Naturally that put new life into me. I always feel excellent when I receive the body and blood of Our Lord."

Roddie prayed and read scripture daily—mostly in solitude. "I want peace, quiet, and more than anything I want God," he wrote in one diary entry.

"Faith was something that you could rely on because . . . [the Germans] did everything that they could do to break me down physically," recalled artilleryman Corporal Russell Gunvalson of the 590th Field Artillery Battalion. "But the only thing I had left was my faith in God. That's one thing they couldn't take away. As a mat-ter of fact, I think it probably made me stronger. It taught me that life is precious and you need something. You can't go through life alone. You've got to have somebody there that you can confide in."

HEIGHTENING THE MISERY of the starvation rations was the in-festation of lice, bedbugs, and other vermin in the barracks. The lice were described by one GI as being "ubiquitous and the size of grains of rice."

All the men, Roddie included, suffered these infestations. The lice collected in the men's armpits and crotches, leaving red bites ultimately all over their bodies. The GIs discovered their own creative ways to "delouse" themselves, even if for just a few hours. Some men chose the flame of a candle, but that required considerable skill and practice to kill the lice without burning holes in your uniform. Beyond painful bites and scratching, as many of the POWs learned, the lice could transmit pyrexia, also known as trench fever. The initial symptoms were shooting pains in the shins coupled with a very high fever. A case of pyrexia could quickly disable any POW; compounded by the freezing conditions and inadequate rations, the lice bites could even cause death.

Eventually, the Nazis themselves became concerned about the health risks and ordered a general delousing.

"Wednesday, we were 'deloused,'" Frankie wrote in his diary. "We were given baths and our clothes were steamed for approximately ½ hour. The bath in itself was useless as we had no soap. But it was pleasant to have our bodies under hot water for a change. That was the day I first noticed how rapidly my weight has been leaving me."

Both Roddie's and Lester's toes had been badly frozen during the fighting in the Schnee Eifel. Those first weeks in the prison camp, they couldn't sleep—the pain was so excruciating they kept doubling over and touching their toes throughout the night, trying to bring them back to life, or at least adjust themselves to the increasing discomfort.

In a war crimes investigation conducted in 1945, many soldiers recounted the numerous cases of dysentery brought on by the malnutrition. The medical supplies were woefully inadequate. "Most of their supplies" were "first aid kits collected from other Prisoners of War, and a few aspirins and other little pills not nearly sufficient enough for the amount of men and sickness. Bandages consisted of crepe paper."

Frankie wrote in his journal about the day the Germans photographed all the POWs: "I can imagine how I must have come out. Since my capture, I hadn't shaved or had my hair cut and I lost considerable weight. I'd never want Lucy or anyone to see those pictures. Hunger pains were becoming unbearable now. I didn't know how any of us could stand it much longer."

For men who smoked, cigarettes were often a more valuable commodity than food.

Frankie wrote: "I traded my watch with an Italian prisoner for one loaf of bread and about 2 lbs of potatoes and about one pound of beet sugar. I shared the food with Roddie, Louie, and Johnny and they in turn gave me quite a sum of money. But I didn't have the money long. The craving for cigarettes became so bad that I began to buy them at the fantastic price of $10.00 per cigarette."

Throughout the day, men could be heard calling out, "Bread for cigarettes! Third of a ration! Bread for cigarettes!"

Roddie, having secreted away all his US cash in his shoulder patch, now put it to good use. He spent most of the money buying cigarettes for guys like Frankie and other friends who were so hooked on nicotine that Roddie could see they were growing desperate.

TWO WEEKS INTO their stay at Stalag IXB, Oberst (Colonel) Karl-Heinz Sieber, the camp commandant, ordered the segregation of American Jewish POWs from non-Jews. All Jews had to identify themselves by six the next morning. Any Jews found in the barracks after this time would be shot. Lester was proud to see that most of the American prisoners banded together and refused to permit the Jewish GIs to obey this first order.

Another soldier recalled a mass refusal from the POWs: "We protested that we were all Americans and wanted to be treated equally, but we were told it was a direct order from the High Command."

Indeed, according to a report made by a Swiss delegation of the International Red Cross on January 24, a Major Siegmann is listed as the "Accompanying Officer from the High Command." Belgian and French military records show that as early as the summer of 1942, when he was only a captain, Siegmann had been with the High Command, specifically tasked with accompanying delegations of the various neutral "protecting powers."

The Oberkommando der Wehrmacht (OKW)—High Command of the German Armed Forces, headed by Field Marshal Wilhelm Keitel and General Alfred Jodl—insisted on a policy of segregating Jewish POWs from non-Jewish POWs. This was in keeping with the overall Nazi policy of anti-Semitic persecution. In September 1941, SS-Obergruppenführer Reinhard Heydrich had ordered that all Jews six years of age and older living in the Nazi Reich be forced to identify themselves in public by wearing a yellow Star of David badge, sewed to the front of their clothing, with the word "Jew" written inside the star.

Following this policy, the OKW issued its own order, on March 3, 1942, pertaining to the treatment and identification of Jewish POWs from Allied nations. This little-known order decreed that captured Jewish soldiers would not have to wear the yellow star, but they *would* be segregated from non-Jewish soldiers in POW camps. This provision had, in the years before the D-Day invasion, been widely applied in the infantry POW camps, primarily affecting captured Jews serving in the French, Belgian, and Dutch armed forces. Only after the Normandy landings in June 1944, and particularly after the capture of thousands of GIs during the Ardennes counteroffensive that winter, did this 1942 OKW order affect significant numbers of Jewish American POWs.

Visiting delegates from various nations observed this religious segregation firsthand and, in the postwar years, offered considerable testimony about it. One delegate for the Vichy French government, Pierre Arnaud, wrote: "We as the delegates always protested [this separation of the Jewish POWs]" into special barracks, or work details, called *kommandos* by the Germans. "Our argument was that the Jewish prisoners were soldiers, like anyone else."

Arnaud observed that French Jewish officers, housed in Oflags, were segregated but not often mistreated. However, in stalags for noncommissioned officers and enlisted men there were no fixed codes of conduct in segregated kommandos, and Jewish prisoners were at the capricious whims of the various commandants and guards.

"I personally visited many such camps," Arnaud testified, noting the degree to which the Germans were "obsessed with race," even in the treatment of captured Allied POWs. Following an official visit to Oflag XC, in Lübeck, he wrote: "I found myself by chance in a barrack where the Nazis had put the Jews and the priests—two groups that were considered by the Germans as the fiercest enemies of Nazism."*

Pierre Arnaud also offers us a fascinating biographical portrait of Major Siegmann. For several years, Arnaud had extensive dealings with the High Command. By and large, he wrote, "we were dealing with officers who were relatively decent. . . . These OKW officers were not always Nazis. However, the most nationalist of them all was a guy named Major Siegmann. He belonged to an old military family and he had served in the war of 1914–1918. Siegmann was a professional career officer." After the end of the First World War, "he saw the social troubles in Germany and was disgusted by the Weimar Republic, so much

---

* Arnaud writes in French that the Germans considered Jews and priests, *"les ennemis les plus farouches de nazism."*

so that he left Germany and moved to the United States," where he rose through the ranks of General Motors as an important executive.

But, Arnaud continues, "when Siegmann saw that war in Europe was again inevitable, he tendered his resignation from General Motors, left his life in the United States and returned to his homeland to volunteer for service in the German Army." He did this out of a sense of patriotism. According to Arnaud's testimony, Siegmann was "very angry at the West," held extreme nationalistic German views, and was a true believer in the Nazi cause.

As the overall head of the Prisoner of War section of the Wehrmacht High Command, Siegmann was essentially the eyes and ears of the OKW in the camps, particularly during the frequent visits of delegations of the various protecting powers. He reported directly to Field Marshal Keitel and General Jodl, both of whom were later convicted of war crimes at the Nuremberg trials and executed by hanging.[*]

Arnaud noted that Major Siegmann "was a very tough and very hard man. When he said 'yes' it was 'yes.' When he said 'no' it was 'no.'"[†]

MAJOR SIEGMANN—ACTING WITH the full weight of the High Command in Berlin—reacted with characteristic harshness and anger to the American Jewish POWs' defiance of the first order. He shortly issued a second order, threatening that "all Jewish violators," when caught, would be summarily shot and that other GIs sheltering the Jews in their barracks would also be executed. The

---

[*] Keitel and Jodl were among the ten most prominent Nazi political and military leaders hanged on October 16, 1946, in the gymnasium of Nuremberg Prison.

[†] Arnaud's testimony in French reads: *"C'était un homme très met y très dur."*

Jewish soldiers met among themselves and, almost to a man, decided to obey rather than subject their comrades to the possibility of such drastic punishment.

LESTER RECALLED HEARING the announcement. "Most of the American troops were with us. I knew my comrades, they were supportive, they didn't want us to do that. When the Germans said they would kill not only the Jews but any soldier that protected them, at that point my friends who were Jewish decided that it was a better part of valor to identify themselves."

Lester, Paul, Skip, and Hank were among the group of Jewish GIs to report forward. They discussed it among themselves and realized that it was senseless to put their Protestant and Catholic buddies—like Roddie and Frankie—at risk. All self-identified Jewish American infantrymen were moved to a barracks surrounded by barbed wire, which Lester described as "a prison within the prison. It was a long building, one story, with beds on either side and straw mattresses. There were all ranks: privates and noncommissioned officers."

No one knew what lay in store for the Jewish GIs, but it must have been a heartrending moment for Roddie and Frankie. The segregation of the men seemed to shock them speechless. It was almost too brazen and chilling to discuss.

Roddie tersely described the event in his diary: "January 18, 1945—Jewish moved out."

Frankie was only slightly more expansive in his entry for that day: "On the 18th of January, the Germans showed their true colors. They segregated all the Jews from the gentiles. I don't know why they did this. . . . After all, they were Americans, and according to the Geneva Treaty, all Americans had to be treated equally."

Though many Jewish GIs, including Lester, Skip, Hank, and

Paul, were already in this "prison within the prison," the Germans knew there were still more Jews among their captives.

Accounts differ on how, exactly, the Jewish POWs at Bad Orb continued to be segregated. Sonny Fox remembered the Germans reading out a list of Jewish prisoners at roll call: "That's a moral crisis for a nineteen-year-old. Do I remain mute, or do I say, 'Take me! I'm Jewish'? Finally, I rationalized that if I was going to help them at all, it would be easier to help them from outside than inside."

That night, Sonny—the supposed "Protestant" sergeant from Brooklyn—was shocked by the anti-Semitic jokes being cracked in the dark barracks by the American soldiers. "It was one of those moments where you think about the distance between who we profess we are as a people and who we really are as individuals [in] society," he recalled.

After a few days, Ernest Kinoy found that he could not live with the guilt of pretending to be Christian while other Jewish GIs were being segregated. "I knew a number of people who were down there in the barracks. I went and turned myself in. Basically, it was an ethical decision, that I didn't think I could separate, get away with something that other people could not. . . . Of course, the possibility of the Germans doing something was apparent, because you don't segregate without something in mind."

Most disturbing of all, at least one American officer in Bad Orb played a role in encouraging the segregation.

Like Roddie and Frankie, Private Sydney Goodman kept a POW diary. On January 18, the day the Jewish soldiers were segregated from the rest of the men, he noted, "All Jewish boys to be separated tonight into where the officers formerly lived. Very cold. I wonder what the future holds in store for us?"

Goodman recalled later what happened to him: "An American officer came in and told us it was the policy of the German government to segregate Jewish soldiers from the rest. He asked all the Jewish soldiers to identify themselves. He said he didn't want to frighten us, but said, 'If some of you guys are Jewish and they find out about it, bad things could happen.' A little later he came back with a list of names that sounded Jewish, and we were all moved to another barracks and locked in."

Private Morton Brimberg, a soldier with the 42nd Infantry, also recalled an American officer urging him to identify himself as a Jew to the Germans. Like almost all the other GIs, Brimberg debated what to do with his *H* dog tags upon capture—he mistakenly thought that no American soldier could be vulnerable to persecution as a Jew. "I tried to say no. But when he asked me a second time I said, 'I am a Jew.'"

A day or two later Brimberg was moved to the hut, which had already been dubbed "the Jewish barracks."

"FORTUNATELY, THE GENEVA CONVENTION requires that prisoners be kept in three separate camps," Lester recalled. "Privates and privates first class in one camp, noncommissioned officers in a separate camp, the commissioned officers in a third camp."

To this day, it remains unclear exactly why the Nazis chose to abide by some provisions of the Geneva Convention at Bad Orb while flagrantly ignoring others. For whatever reason, the Nazis had decided to comply with the regulation requiring the differentiation of noncommissioned officers from privates and privates first class. A precipitating event was a visit from the International Red Cross on January 24, 1945. A delegate of the Protecting Power (which for the US in World War II was the neutral country Switzerland) found many of the men enduring dreadful overcrowding—sleeping on bare floors, without blankets, and with inadequate heating. The delegate had heard of multiple deaths from diseases already among the POWs.

The official report from the Red Cross reads:

The Delegate of the Protecting Power told the Camp Commander that in his opinion, the present accommodation was untenable. In reply, the Camp Commander claimed that this is only a transit camp and that the prisoners would be transferred elsewhere at the earliest possible opportunity. However, in the view of the present situation in Germany, an early transfer is unlikely, and the Delegate of the Protecting Power insisted that improvements in the accommodation should be made at once, in case prisoners should be kept here after all. At this point the Camp Commander informed the Delegate the N.C.O.'s mentioned above were to be transferred elsewhere the following day.

A sinister new character soon arrived in Stalag IXB: SS-Untersturmführer Willy Hack. Hack was supervising slave labor at a sub-camp of Buchenwald. He needed at least 350 more prisoners from Stalag IXB for a "special work detail." The SS had "full authority" to take POWs under the rank of noncommissioned officers for forced labor.

German regulations stated that "Jewish prisoners of war may be grouped in closed units for work outside the camp."

Hack and Sieber knew that there were still more Jews who had not identified themselves within the American POWs, and by all accounts, they were determined to find them.

AT DAWN ON January 25, Sieber ordered all the noncoms—Jews and non-Jews alike—to stand in formation for hours in subfreezing conditions. The snow was knee-deep. The wind was howling. Accounts vary slightly, but by best estimates there were a total of 1,292 American noncoms at Stalag IXB.

Finally, after more than four hours of standing in the cold, the men were marched downhill toward the Bad Orb train station. By that time, most of the men were limping along with frozen feet.

They were each issued one-third of a loaf of the black sawdust bread and one-third of a can of bully beef, then they were told they were bound for a new camp. "In spite of the bitter cold and our frostbitten feet, our spirits were much higher. We had hope of better food and some long-prayed-for Red Cross packages," Frankie wrote in his journal.

After the two-hour march to the train station, they were once again placed in boxcars before starting the long journey to a new camp.

The train traveled some 100 miles northwest, to Stalag IXA, a camp for noncom infantrymen, located in Ziegenhain, a small town in the Rhineland-Palatinate.

When the door of their boxcars opened, Roddie saw they'd arrived at a railroad marshaling yard with multiple tracks and switches that had been recently bombed by the Allies. It was a dreadful march to the camp gates in subfreezing temperatures. On the way from the station, the Americans found themselves facing nearly blinding blizzard conditions. Wind whipped across the barren fields surrounding them, but most men kept their heads down, tucked into their jacket collars, keeping their gaze on the boots of the GI marching in front of them.

"Everyone trudged along in his own individual world of pain," as Richard Peterson recalled.

By now Roddie and the other GIs had been surviving for more than a month on less than 500 calories per day, and most had already lost 20 pounds. Their physical deterioration in those weeks of captivity was startling. They'd been wearing the same clothes in which they had been captured; no one had changed underwear or socks. They were infested with lice and stricken with dysentery. Most of the infantrymen—eighteen- and nineteen-year-old boys—were suffering from frostbite and trench foot. Roddie's feet were in excruciating pain.

The German guards had no mercy. Throughout the grueling march they shouted for the men not to fall behind or slow their pace at the risk of a gunshot.

Inside the gates of Stalag IXA, the men were forced to stand in formation for hours while large German shepherds—trained by the camp *Hundführer* to be ferocious—growled menacingly. Finally, just before sunset two German guards brought forward a young Soviet POW. They were joined by their new camp commandant, Oberst Hans Mangelsdorf, and his adjutant, Hauptmann (Captain) Fritz Bock. The Soviet POW looked to be no more than eighteen or nineteen. He was gaunt, on the verge of starvation, unshaved, and wild-eyed.

"You're free to go," Mangelsdorf announced. The Russian POW, realizing it must be a ruse, refused to budge. The Germans ordered that the camp gates be opened. The Russian POW saw no option; with rifles pointed at him, he took a few tentative steps, then began running toward the gates.

The commandant motioned with his hand, and the *Hundführer* unleashed the dogs.

Within seconds, the Russian was mauled, falling to his knees, screaming, until teeth finally clamped on his throat.

At last the boy's screams stopped.

My father's diary entry for that day was a single word, direct and to the point, though it took me years to decipher its meaning, to appreciate in full the weight of his experience. My father simply wrote "Dogs."

"Remember this well," Mangelsdorf told the newly arrived Americans. "If any one of you disobeys orders, the same fate awaits you."

NINETEEN

P AUL STERN COULDN'T stop thinking about the boy without a face.

Back in the early hours of the fighting in the Ardennes, on the afternoon of December 16, Paul had been scrambling among the bodies of the wounded, the dying, and the dead. He had rushed to the side of a young GI who had stepped on a German mine. The soldier was only nineteen years old, just like Paul and the other field medics.

Paul tried his best to treat him, but the kid's face was blown off. His face was *gone*, but he was somehow still alive.

Paul had never felt so powerless as a combat medic. How was the young GI even breathing? How was he conscious? All Paul could do was sprinkle a little sulfanilamide on the wounds where the boy's face should have been—Paul's unit had no penicillin or morphine.

Bobby—yes, that was the boy's name. Paul was sure he hadn't lived more than a day. Paul had prayed for him and had gotten him evacuated to the rear lines, but . . . that picture stayed in his mind, haunting him throughout his nights in Stalag IXA.

A day never went by when Paul didn't think about Bobby.

And a day never went by when he didn't think about the young lieutenant and the other soldiers standing by a pillbox in the Siegfried Line.

In the first days of the fighting in the Ardennes, a frightened-looking lieutenant had approached Paul for advice in a rest area. He was a medical administrative officer, had never seen a moment of combat, had never fired a gun in battle, and had been promoted to infantry leader without having had much training.

"Stern," he said, "you've been around quite a bit—I don't know a thing about combat. Can you help me? What do I do?"

Paul was taken aback. What could he possibly tell the lieutenant? "All I can tell you is to keep your butt as close to the ground as possible and pray like hell. That's the only thing you can do when you're under fire, sir."

Just moments later, he looked up and saw three GIs gathered around the entrance to a German pillbox. They were standing there, leaning right up against the pillbox, relaxed, smoking cigarettes, completely exposed.

Paul knew there were Germans inside the pillbox, and he sprinted as fast as he could—some fifty yards. As soon as Paul got close, panting, he recognized the same green lieutenant who'd asked for his advice. That lieutenant and the two other soldiers were sitting ducks.

"You guys gotta move!" Paul shouted, pulling the three men to the south side of the pillbox—concrete so thick that no bullets or shrapnel could penetrate it.

He pulled them out of the way just in time. Moments later, a German soldier inside the pillbox opened a slit in the door and tossed out a hand grenade. It exploded right in the spot where the lieutenant and soldiers had been standing. They recoiled from the blast, unhurt behind the concrete wall. And they hurried back to the safety of the battalion aid station.

The lieutenant wasted no time in telling Paul's commanding officer about the incident. The next morning, Paul was promoted in the field, bumped up from private to corporal, a noncommissioned officer. He also received the Bronze Star for heroism.

Lying in the barracks in Stalag IXA with the other noncoms, Paul fully grasped what had happened at that German pillbox: he'd saved three men's lives, and by doing so, without realizing it, they might have saved his. He didn't know what lay in store for the Jewish privates and privates first class back at Bad Orb, but at least, as a noncom now, Corporal Paul Stern could be fairly certain he wouldn't be separated from his close friends Staff Sergeant Lester Tannenbaum and Corporal Skip Friedman.

Paul would never forget that lesson. "Whenever you help people," he often said later, "you help yourself."

PAUL AND SKIP repeatedly lifted each other's spirits—even as they saw their bodies deteriorate with starvation. Several times Paul, ever the optimist, made a promise to Skip that they would be free men by Passover that spring. It surely could not have been a coincidence that Paul's Hebrew name was Pesach (or "Passover").

"Paul and I found a way of keeping each other 'up' in comparison to the other GIs," Skip later recalled. "He assured me we would be eating matzoh by Passover."

Paul was a young man who never lost faith. He'd already witnessed several miraculous episodes during his service in Europe. In the days before the horrific Battle of Hürtgen Forest, in late October, Paul had arrived in the ancient city of Aachen—the first German city to be captured by the Allies.

And on Rosh Hashanah, the Jewish New Year, an extraordinary ceremony had been arranged. Paul gathered with more than fifty other Jewish GIs from various combat units to conduct the Jewish services. Several of the men draped their shoulders with prayer shawls. Many, like Paul, remained in full combat gear, bowing their heads in prayer while still wearing their steel helmets. On October 29, 1944, amid the pillboxes of the Siegfried Line, and near an old Jewish cemetery, radio microphones and cameras captured the historic moment: a proud gathering of GIs, singing and reciting Hebrew prayers.

The fighting was so close that, as they prayed, the Jewish GIs could hear artillery shells exploding. The medieval synagogue of Aachen had been destroyed years before—it was among the hundreds of Jewish temples attacked by rampaging mobs on Kristallnacht in 1938. But then, even though the synagogue was a grim ruin and many of the city's Jewish citizens had been murdered, the words of the Jewish New Year service were heard again in Aachen: "For a thousand years in thy sight are but as yesterday when it is past, and as a watch in the night."

The Rosh Hashanah service was broadcast the next day on the NBC radio network throughout the United States. And later, the service was broadcast into Nazi Germany. It was the first time,

since the rise of Hitler, that Hebrew prayers had been heard over German airwaves.

The day after the ceremony, Paul had returned to the cemetery with a Jewish American lieutenant. In a large tomb they found hundreds of rare Jewish religious books and Torahs, which, they reasoned, had been placed there by Aachen's Jewish leaders for safekeeping before the synagogue had been set ablaze. Paul stood amazed—many of the books were so fragile, they had to be eight hundred or nine hundred years old, a priceless library that Paul and the lieutenant were concerned might be destroyed in the shelling.

Across the street was a Roman Catholic church and monastery, and Paul asked the Mother Superior if she would take all the religious books and hide them in the church.

"If the Nazis ever come back," she said in a whisper, "they will cut all of our heads off."

But the Mother Superior was brave; she said, "Yes."

It took Paul and the lieutenant half a day, but they managed to safely hide the sacred ancient Torahs and prayer books in the church.

In the barracks at Stalag IXA, Paul was suffering, like all the other men, from the effects of starvation. But he never lost faith. He told Skip over and over that he was certain they would survive. "Skip," Paul vowed, "we're going to celebrate Passover as free men."

Still, even as he entertained thoughts about his life as a free man, he couldn't escape the image of Bobby or his increasing concern about the Jewish soldiers left behind at Bad Orb.

AT STALAG IXB SS-Untersturmführer Willy Hack was intent on filling his work contingent of 350 Americans—Jewish or otherwise—and began to interrogate other soldiers. Not only men with clearly

Jewish surnames, like Private William Shapiro, a medic from the Bronx, but some who simply had "Jewish-looking" faces. Or "Jewish-sounding" names. Hack filled the quota with non-Jewish "troublemakers" and "undesirables"—like Private First Class Johann Carl Friedrich Kasten.

When officers like Sieber and Hack continued to press—under threat of execution—for the remaining Jewish GIs, Kasten protested. Kasten was a *Vertrauensmann*, a "man of confidence," one of fifteen men of confidence the camp had elected to represent the enlisted soldiers before the camp's commandant. "This was for the confidence of our men not the Germans," Kasten recalled. "I was the unlucky one to be elected." His fluency in German made him a natural choice.

Born and raised in Hawaii, Kasten was a second-generation German American. He had visited prewar Nazi Germany as a teenager, even personally met Adolf Hitler, but he was nonetheless a patriotic American who had enlisted in the US Infantry in 1943. He had landed on Omaha Beach during D-Day and had been captured, during the Battle of the Bulge, fighting with the 28th Infantry Division.

Kasten was summoned to a small second-floor conference room, where several German officers were waiting.

"There were eight chairs and [on a table] in front of one of the chairs was a loaf of bread—very obviously a bribe for something coming," he recalled.

He took his seat, and the senior German officer demanded, "Kasten, we want the names of all the Jews in the American camp."

Without any hesitation Kasten pushed the loaf of bread to the center of the table. "We are all Americans. We don't differentiate by religion."

Kasten was jerked out of his chair and thrown down a flight of stairs from the second floor. He lay in the street, "trying to determine if any bones were broken," then when he was finally able to stand, badly bruised, he returned to the American sector of the camp and immediately summoned the other men of confidence. He instructed all the leaders to return to their barracks and relay the story, with his final instructions: "Something is bound to happen and soon, and none of the men should admit to being Jewish."

Sure enough, that afternoon the Germans ordered the entire camp out onto the parade ground. There were some four thousand US infantrymen with the fifteen men of confidence in the front.

The commandant stood on a small platform and shouted: *"Alles Juden, ein Schritt vorwärts!"* ("All Jews, one step forward!")

"Because of my morning instructions no one moved," Kasten recalled. "This infuriated the officer to an extent that he jumped down from his platform, grabbed a rifle from one of the guards, and rushed towards me. I was convinced he was going to shoot me, but instead, holding the barrel, he swung the rifle like Babe Ruth and with all his strength crashed the butt against my chest. I flew backwards about fifteen feet and fell on the ground. I couldn't breathe. I thought I was done for."

While Kasten lay there dazed, the guards went through the ranks of men and pulled out any GIs who "looked Jewish."

"Kasten, Sie und Ihren beiden Assistenten auch." ("Kasten, you and your two assistants also.")

Kasten felt this was retribution for his having been frank with the inspectors from the Red Cross about the brutality and conditions at Bad Orb as well as his refusal to turn over Jewish troops.

Hack also took a twenty-year-old Mexican American from Southern California, a combat medic with the 275th Infantry Regiment, Private Tony Acevedo. Hack accused Acevedo of "lying and spying against Germany" in the northern Mexican state of Durango, even though Acevedo was from San Bernardino, California, and had nothing to do with counterintelligence. Convinced that Acevedo was a spy, Hack ordered Acevedo's torture. "They put needles in my fingernails," Acevedo later recalled. "Like nailing me to the cross." Still, Acevedo would not lie, would not falsely confess. He was segregated with the men bound for the "special work detail."

The GIs who were taken from Bad Orb, Kasten later recounted, were "crammed into box cars as before—no food or water—four days to the town, Berga an der Elster." The POWs soon saw that "Berga was a slave labor concentration camp. Totally against Geneva Convention rules for prisoners of war to be incarcerated in such a place. The Jewish men selected [Morton] Goldstein as their leader, but the Germans told them a Jew could not hold that position, so automatically I became the leader as a carryover from Stalag IXB. Very rapidly things went from bad to worse."

The slave-labor detail at Berga, near Dresden—comprised mostly of political prisoners and European Jews from concentration camps—was part of an "emergency fuel program," established on May 30, 1944, to transfer synthetic oil production deep underground in order to protect it from Allied bombing. Known as Schwalbe V, the top-secret project was under the immediate direction of Hack and entailed a massive construction zone of seventeen different tunnels leading into a planned large factory area.

Everything in Berga was under the supervision of the SS. Overnight, American infantrymen from the 28th and 106th Infantry Divisions—including Ernie Kinoy—found themselves alongside prisoners from the Buchenwald concentration camp. In freezing and inhumane conditions, the GIs were forced to blast and hack out the tunnels, all the while inhaling large dust particulates, which left most of them with lung ailments. Inside the Berga concentration camp, slave-laborers endured constant mistreatment, starvation, and beatings by Willie Hack and his second-in-command, a fanatical SS guard named Erwin Metz.

Their only meal was a weak soup made from weeds mixed with a dead cat or rat, and a tiny portion of "bread" containing sawdust, ground grass, and sand—only 100 grams of it per week. The European-Jewish workers from Buchenwald were killed frequently for minor infractions or at the capricious whims of the SS, and the American GIs were forced to watch the SS men execute them by beating and hanging.

Under Hack's sadistic watch, the Americans were now also forced to be slave-laborers, alongside European Jews from concentration camps, on the massive demolition project, tasked with digging deep tunnels into the mountains along the bank of the White Elster River.

"The agony of it all soon became evident," Kasten later wrote. "We had to get up while it was still dark, march to the mines to dig the tunnels without food or water and 12 hours breathing stone dust and hauling out carts loaded with stone." Very soon after arrival, "the men started dying of exhaustion, malnutrition, and extremely harsh treatment."

Private William Shapiro was—as a combat medic—spared from work in the mines but watched the gruesome and often fatal slave-labor details each day. The work went on "seven days per week," he wrote. "They were marched to the tunnel area about a mile from our barracks by our guards. In the slave tunnels they were under the command of the same work gang bosses and civilian engineers as the slave political prisoners from the concentration camp annex. Their work was identical to the political prisoners but at a different shift time . . .

"The slave work consisted of excavating rocks and dirt by hand and shovel after it was loosened by the explosives. The men hand loaded rocks onto or shoveled slate fragments and dirt onto flatbed cars similar to coal cars. They hand pushed the cars on its track to

an area where the rocks could be dumped into the [White] Elster
River. They worked with primitive drills, old mining machines
and often the men were utilized in place of machine power or
horse power to move heavy objects."

Fatal accidents and brutal beatings by the guards, armed with
rubber hoses, were commonplace.

"Slate dust was choking and ever-present," Shapiro said. "Our
Volkssturm guards marched the shifts of POWs to their tunnels,
where they remained until the shift ended. The return to the barracks was often marred by the indifference and uncaring long
waiting periods until our guards arrived to march them back to
the barracks and their bunks. Even after this torture there was a
further delay of the much-deserved rest because they had to stand
in line for the evening count of the prisoners. These repetitive,
disruptive, inane counts made the suffering harsher. Finally, the
meager rations of a bowl of rotted potato or turnip top green soup
and a slice of hard, grainy black bread was distributed."

With his medical training, Shapiro did his best to tend to the
sick and starving GIs—though he had no medical supplies.

"It was very apparent to observe their progressive bodily deterioration and the loss of will to continue to live," he wrote. "It
was in their vacant staring into space, the exhausted movements
and disinterest in their surroundings which foretold the ultimate
demise of many of them." Even when not being tormented by the
SS men, Shapiro observed, "our agony was worsened by supposed
ordinary Germans."

Many of the overseers were men of the Volkssturm, some as
old as seventy. These were "grown men of age when Hitler came
to power in Germany. . . . They were older men who possessed
the same mentality of the SS even though they were the last resort
of men to defend their country as Home Guard. They were con-
scripted for guard duty because of age and disabilities . . . yet their

actions were equally barbaric as any Nazi. They did not treat us as soldiers but rather as political prisoners who had to suffer. . . . A large measure of their brutality was [of] their own volition as they were not under constant orders or surveillance by their superior officers. Beating the prisoners, hitting them with the butts of their rifles were spontaneous expressions of their own individual bestial acts against mankind."

Then there were the hangings of the European Jews in their telltale concentration camp uniforms with the yellow stars. "The bodies of European Jews, emaciated in their striped uniforms, were left suspended from a robust crossbeam as warnings to other inmates," author Roger Cohen wrote.

Ernie Kinoy thought he'd seen hellish scenes before—in the Battle of the Bulge, in the forced march, and at Bad Orb. But it was nothing like the scenes at Berga. One winter day Kinoy was selected by the SS men to leave the mines and go to the Berga concentration camp for supplies, then drag them back up the hill. It was back-breaking labor, and he found himself standing in a yard as a Nazi sergeant started to beat a young Hungarian Jewish prisoner with a whip: "Across the street were three or four German women with baby carriages," Kinoy recalled. "The guard would scream at this 13-year-old kid, call him to attention, then viciously beat him again. And you could see these three or four ladies from the town standing there, continuing their discussion, watching this Nazi beat up the kid. And I've always found that significant as a symbol in terms of what the German public did and did not know about what was going on [in the Holocaust]. These ladies watched it all and did not even once turn away."

FOR THE GIS condemned to Berga, death seemed almost assured. In time they resigned themselves to it. Others seemed unbreakable. Like Kasten.

Undaunted by the brutality, Kasten wrote that he "requested a meeting with Lt. Hack, accompanied by my assistant Joe Littell. We made our point about the Geneva Convention rules and that one day he would have to answer for his actions."

Hack's answer was blunt: "You were brought here to work and that's what you will do!"

Hack mockingly read aloud the GI's full name from his ID tags: *Johann Carl Friedrich Kasten.*

"You are a German and you have come here to destroy the Third Reich. You know, Kasten, there is only one thing worse than a Jew. Do you know what that is?"

Kasten remained defiantly silent.

"It's a traitor who betrays his country. And you are a traitor."

Hack turned his ferocious gaze to Private Joe Littell.

"Are you Jewish?" Hack demanded.

"No, I'm a Christian," Littell said. "I'm a follower of Jesus Christ—and Jesus was a Jew."

"Leave Jesus out of this!" Hack shouted.

WILLIAM SHAPIRO SPENT almost every morning on breakfast detail, wheeling a heavy cart uphill and through the entrance of the concentration camp, where European Jewish prisoners would ladle out the soup. "On very harsh blustery days, they would assign ten men to the food detail. Adding more men to the detail was an opportunity in which we were able to assist some of our buddies in their escape from the camp. We would try to confuse the guards by telling them that we only had eight men on the detail at the start but, in fact, we were ten men. We would move about changing positions on the wagon to confuse the guards into believing that there were eight of us. During the darkness of the morning, beyond the lights of our camp, they would slip away from the detail. Several men escaped the camp this way."

Once, on early morning food patrol, Kasten, Littell, and another fluent German-speaking GI broke away from the guards. They were quickly recaptured and sent to Stalag IXC, where they were held in solitary confinement.

Private Morton Goldstein of the 106th Division kept vowing to escape Nazi captivity. "I had met him in Stalag IXB and he was a garrulous, bombastic man who could not be confined and certainly could not do the work in the tunnels," Shapiro recalled. "His first escape ended in recapture and extra duty, and he was forced to stand out in the cold for a long time."

Around the third week of March, Goldstein made a second escape attempt. He was on the loose only for a few days before being recaptured. Sergeant Metz stood Goldstein against a wall and shot him, execution style, through the head. As his body lay on the ground in front of the other prisoners, German guards riddled him with bullets so that he could be classified as "shot while trying to escape." Afterward, the Nazis would not allow his corpse to be moved. "His body laid on a stretcher between two barracks as a warning to others," Shapiro wrote. "The display of Goldstein's body was to deter any further attempts to escape."

The horrors of slave labor in Berga continued for weeks, even when it became clear to the Nazi guards the war was likely lost. Berga was virtually the only POW camp in which American soldiers experienced brutality commensurate with Nazi concentration and death camps. Combat medic Tony Acevedo, who had weighed 149 pounds upon capture, dropped down to a skeletal 87 pounds.

Of the 350 young GIs sent to work at Berga, 73 men died in the space of 10 weeks. The fatality rate at Berga would prove to be the highest of any camp where American POWs were held during World War II.

TWENTY

S TALAG IXA, WHERE Roddie and the other men had been transferred, was a sprawling older camp carefully divided into sub-
camps by nationality: the French, who'd been POWs since the fall
of France in 1940; the British, many of whom had been captured at
Dunkirk; the Serbians; the Soviet Red Army troops from the Eastern Front, treated much more brutally than any other captured
nationality; and now the Americans, the newest arrivals, who'd
been captured by the thousands since the Battle of the Bulge.

The stalag was strictly for noncoms of various nations, and Roddie and the other GIs would quickly learn that the Germans had
forced the defeated French to build it immediately after the Battle of France. Many of the French infantry sergeants and corporals
had been prisoners for five full years. Among these early POWs had
been a young sergeant injured and captured by the Germans on
June 14, 1940, François Mitterrand—who, after the war, would be
elected president of the French Republic. (By the time Roddie and
the other transferred POWs arrived from Bad Orb, Mitterrand had
escaped from Stalag IXA to join the Resistance.)

The captured troops of the Forces Armées Françaises constituted the largest cohort of POWs in Stalag IXA. Given their five
years in captivity, the French had developed a very well-organized
camp "civilization" with a library, sports teams, arts clubs, live
theater, a symphony, jazz orchestras, a choir, even a "temporary
university" where various POWs delivered a series of lectures.
The French had a functioning Roman Catholic Church and six or
seven priests among their POWs.

By this late stage of the war the American prisoners would have
no access to such luxuries as musical instruments or sports equipment; nor would they receive sufficient food—not even Red Cross
parcels—to have the energy to be physically active.

All the Americans could see the French handwriting—
literally—in their barracks. On the filthy walls were the phrases
Defense de fumer ("No smoking") and *Ferme la porte* ("Close the
door"). As one American POW observed, Stalag IXA was "a
French camp in every sense of the word."

The camp was laid out almost like a small town, with buildings
in a line on either side of a central street. On days when the ice
thawed, the street turned to muddy slop. But just as at Bad Orb,
barbed wire was everywhere.

As master sergeant, Roddie Edmonds was now the top-ranking
infantryman in the camp. He was the noncom with the most se-
niority, due to his enlistment date in the peacetime 1941 army. But
it was also evident to all the other corporals and sergeants that he
knew how to look out for their best interests. "Roddie by now was
our barracks leader," Frankie wrote in his diary. "He got the job
because he knew how to give good commands and he was a good
soldier."

Skip recalled, "Roddie was a very stoic guy, very solid guy, and
would take no garbage from anybody—particularly Germans.
We were very lucky to have him with us."

Each of the five American barracks in Stalag IXA had between 250 and 300 men. The overcrowding was not as bad as at Stalag IXB, but the barracks were equally as grim: infested with vermin, dirty, dark, freezing, and barely habitable. There was a cold-water tap, no toilet in the barracks, and a small stove that rarely had any fuel. Each man was given a thin blanket, but they still huddled together for warmth. Starving infantrymen crowded into the triple-tiered bunks, sleeping on thin burlap-sack mattresses filled with straw. Two or three men shared each narrow bunk to keep from freezing during the night.

FIRST THING IN THE MORNING, the day after the Americans were transferred, the staff car of a Nazi officer arrived in Stalag IXA. Word quickly spread that it was Major Siegmann, from the Oberkommando der Wehrmacht, who had followed the noncoms from Bad Orb to Ziegenhain.

"*Achtung!*"

In the late afternoon the camp loudspeakers boomed with those two guttural syllables, shattering the frozen silence of the stalag. Then, after a long crackling pause, came the order, first in German, then in English:

Tomorrow morning at roll call all Jewish Americans must assemble in the Appelplatz [the place where roll call is performed]—only the Jews—no one else. All who disobey this order will be shot.

Roddie listened closely along with Frankie, Lester, and the others in their barracks.

"These were the same orders we'd received at Bad Orb," Lester recalled. "Only this time, we were organized. Roddie, for the first time in this experience, was in complete command. There was no one there to give him orders. It was his decision."

Without hesitation, Roddie turned to his men and said, "We're not doing that. Tomorrow we all fall out just as we do every morning."

Then Roddie was silent. Frankie knew that Roddie would never comply with an illegal and immoral order—he was too good a soldier and too decisive a leader. But how could Roddie defy the Germans without putting everyone's lives at risk?

Roddie was deep in thought. He was most likely praying and recalling a favorite Bible verse—as he often did in life. I know that his favorite passage in all of scripture was Romans 8:37:

Yet in all these things we are more than conquerors through him who loved us.

How could 1,292 malnourished, frostbitten men suddenly be transformed into conquering lions? Any defiant action he might take could be potentially deadly—not just for the Jewish GIs but for every single infantryman in the camp.

There was a simple solution—but risky. Roddie called a meeting of all barracks leaders. Several of the most senior sergeants gathered around Roddie's bunk as three other noncoms stood as lookouts at the door and windows.

"We're not doing it," Roddie said.

Every infantryman, he told them, would assemble in strict military formation at the Appelplatz at the next morning's roll call. Every soldier—even those named McCoy and Walker, Smith and Nicholson, Miller and Bruno—would tell the Germans that they were Jewish. Roddie made clear that everyone must follow his order: all infantrymen—every *single* one of the 1,292 men in camp. He stressed that even the men too sick and weak to walk could not be left behind in the barracks. Every man must assemble in the Appelplatz. Because the stakes were so high, Roddie couldn't afford any misunderstanding or inadvertent acts of disobedience. He ordered all the barracks leaders to make sure every man in the camp understood the plan.

After nightfall, once the windows of the American barracks were shuttered, Lester awoke with a start. The words of the loudspeaker announcement kept replaying in his mind. *Tomorrow morning at roll call all Jewish Americans must assemble in the Appelplatz—only the Jews—no one else. All who disobey this order will be shot.*

A few bits of straw, dislodged from the mattress above him, fell slowly like snowflakes and rested on his hollowed chest. He flicked the straw away with the back of his hand. Better than the lice, he thought, that hopped constantly between their bodies.

It was still nearly pitch-black inside the barracks and felt like ten degrees below freezing.

He cupped the steam of his warm breath in his numb fingers, then looked over at Roddie, barely two feet away: in the dim light he could see that his master sergeant was quietly praying.

The wicked flee when no man pursues, Roddie prayed. *But the righteous are bold as a lion.*

Lester watched as Frankie, pressed tight against Roddie, also started stirring. None of the three sergeants could sleep. Frankie's shock of thick black curly hair had grown wilder and matted during recent weeks. Short but once powerfully muscled, Frankie was by now emaciated, his cheekbones jutting. He had lost 30 pounds since the capture after the ferocious fighting around Saint Vith.

Roddie gripped Frankie lightly, perhaps feeling his bunkmate's shoulder bone pressing hard against the fabric of his filthy fatigues.

"You're positive everyone knows, Frankie?" Lester could hear the urgency in Roddie's voice.

"Yeah," Frankie said.

"All five barracks?"

"Everyone, yeah. I told them all again—all other staff sergeants, all barracks chiefs."

"And the guys too sick to walk . . ."

Lester saw Frankie nodding resolutely. "Even the guys too sick to walk."

ON JANUARY 27, at precisely 0600—with the bark of *"Raus! Raus!"*—there was a sudden banging on the barracks doors.

Frankie and Lester hustled the men outside. The hard-packed snow crunched under their shuffling boots. Roddie glanced back to see if *all* the men were there as the barracks leaders had promised. Standing next to him were Tannenbaum and Stern. Nearby were Cerenzia and Friedman.

The men were assembling as planned. Even those too sick to walk were doing their best to stand up straight in formation. A few were having trouble, leaning heavily on other POWs' shoulders—but they were forming up in ranks.

Roddie had faith. Faith in God, faith in his fellow infantrymen. He knew they wouldn't betray their brothers-in-arms, not voluntarily. But he *also* knew that starvation, and desperation, could play games with men's minds. He was starving and exhausted, in terrible pain, nearly physically and emotionally beaten. He started to worry about his and his men's resolve. What if a single starving soldier decided to save his own skin by pointing out the Jews in the ranks? What if the plan didn't work?

He tried to steel himself. One of his favorite scripture passages, which he loved to quote throughout his life, was John 15:13:

Greater love hath no man than this, that a man lay down his life for his friends.

SUDDENLY, MAJOR SIEGMANN approached the Appelplatz.

As he got closer, Roddie could see his face clearly, his blue eyes scanning the ranks of men in the Appelplatz, his black polished boots gleaming in the morning light.

"*Was ist los? Ist das ein Witz?*"

A joke? The major no doubt had expected to see a smattering of Jewish GIs—or maybe a couple hundred, just like the selection back at Bad Orb—and could scarcely believe his eyes. The entire American camp—all 1,292 US prisoners—stood lined up in sharp formation.

He stormed directly toward Roddie and shouted in English. "What's this?"

Roddie held his strict posture, jaw fixed, looking straight ahead. "Under Article Seventeen of the Geneva Convention," he told Siegmann, "prisoners of war are only required to provide name, rank, and serial number."

The major approached to within striking distance of Roddie. "Were my orders not *clear*, Sergeant? Only the Jews were to fall out."

On his immediate left, Roddie could see Lester, and on his right, Paul. Though they were both clearly terrified, like Roddie, they kept up stoic expressions.

"Major, we'll give you name, rank, and serial number," Roddie said. "That's all."

"Only the Jews!" Siegmann shouted. "They cannot all be Jews."

Roddie turned to stare the major directly in the eyes. "We are all Jews here," Roddie said.

Roddie's defiance spread throughout the ranks. A sense of unified strength coursed through the starving, weakened POWs, who at this point barely resembled soldiers. Hearing their leader's calm resolve emboldened them again. Roddie's words and actions seemed to give *every* infantryman in the camp courage.

Not a single soldier broke ranks, faltered, or flinched.

Stepping forward, Siegmann drew his Luger from his holster.

"Sergeant, one last chance," he said, as he pressed the barrel of the pistol hard against Roddie's forehead, right between his eyes. "You will order the Jews to step forward or I will shoot you right now."

Roddie stood his ground.

There was a long silence between them. No words. No gestures. Only the swirling gusts of snow and the smoke-like puffs of their frozen breath dissipating skyward.

At last Roddie replied, calmly. "Major, you can shoot me, but you'll have to kill *all* of us—because we know who you are—and you *will* be tried for war crimes when we win this war. And you *will* pay."

The major's face blanched; his arm trembled.

The Luger was still pressed to Roddie's head—his finger still on the trigger.

Then, quickly—enraged—Siegmann snapped the pistol back to his side, holstered it, turned on his boot heel, and fled the compound.

PART V

The opposite of love is not hate, it's indifference.

—ELIE WIESEL

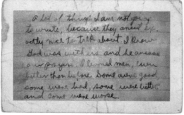

TWENTY-ONE

LESTER RECALLED THE mood of elation in the barracks after Roddie's confrontation with Siegmann: "We went back to the barracks and really cheered Roddie. He never wavered. What he did made us brave. I was very proud—and happy—but Roddie didn't want to talk about it. He was facing the enemy at the risk of his life, and what he did was to save us. He thought only of his men. That takes *courage*. But that was Roddie. We all came to admire and respect him—so that when Roddie said something while we were prisoners, that was what we did."

But all the POWs, Jewish and non-Jewish, remained in constant peril. Every day brought the risk of death: freezing, starving, being shot by a German guard for some minor infraction. The fear was ever-present that the High Command would make yet another attempt to remove the Jews.

Roddie realized the desperation of the moment. And he recognized that his most immediate challenge was to restore a sense of military discipline to the ranks of GIs in Stalag IXA.

He had seen the demoralization and despair caused at Bad Orb. "The way we were mixed in rank and representing three or four infantry divisions, there wasn't much organization. To tell you the truth, military discipline sort of went to hell in that first prison," recalled Pete Frampton. "Our officers were being treated like GIs and given no way to exert any influence for our betterment. The non-commissioned officers were without authority and this led to a certain degree of slovenliness."

In a page of his diary, Frankie Cerenzia wrote of the deteriorating morale and behavior among the American POWs: "Only a small handful have managed to hang on to their pride and self-respect." Many of the men had "stooped to everything but murder over a spoonful of soup or a crumb of bread. Stealing is simply terrible and if you manage to save anything for the next day, you've got to sleep on it in order to hang on to it that long. It's awful to see a once carefree and reckless American hunk of humanity stoop to a form of animal. I pray God to keep me strong in mind and body and I thank him for keeping me this long."

In late January, Roddie realized that, in addition to faith and prayer, the noncommissioned officers would need a new sense of military discipline—to keep the men sane, to curb the cutthroat survival-at-all-cost impulses of some of the more desperate among them.

The 250 to 300 men in each barracks organized themselves into groups of 50, with the highest-ranking sergeant as the leader, and Roddie in overall command as the ranking noncom. They took the initiative to create a staff consisting of six master and first sergeants. Each of the six was given an area of responsibility: health and sanitation, food supply, military discipline, and maintaining a chain of command.

Lester said he knew by heart the precise and detailed regulations for soldiers taken prisoner by the enemy, which are taught at the US Army Infantry School in Fort Benning. Roddie was also well acquainted with those rules:

If I am captured, I remain a soldier. I am guided by the Code of Conduct and subject to the Uniform Code of Military Justice. I am entitled to protection under the provisions of the Geneva Convention. . . . I will take no part in any action which might be harmful to my comrades. And more to this point—I will make every effort to escape and aid others to escape. I must be prepared to take advantage of escape opportunities whenever they arise. POWs must organize in a military manner under the senior person eligible for command and the senior person shall assume command. Strong leadership is essential to survival, and non-commissioned officers will continue to carry out their responsibilities and exercise their authority in captivity.

Camaraderie and faith were crucial, but leaders like Roddie and the other ranking noncoms realized there was also a need for intellectual stimulation.

"For the most part, we were almost as bored as we were hungry," Sonny Fox wrote. The men organized a camp lecture circuit; Sonny and four other "impresarios" took an "inventory" among the POWs to see who could speak on intellectually stimulating topics. "We turned up a ranger from Yellowstone National Park and the editor of the *Bisbee* (AZ) *Bee*. We organized a speaker's tour from barrack to barrack. We would announce the schedule in the barracks and pick a location among the bunk beds for each discourse. The lectures attracted any number from a few to as many as a hundred."

In a camp of nearly 1,300 noncoms, there were young infantrymen with diverse experiences from their civilian lives. "One man
had worked for the district attorney's office in New York City and
regaled us for many days about the mob activities there," recalled
Richard Peterson. "It was so interesting that he must have been
giving us classified information." In the camp, by one soldier's account, there was also a New York Broadway producer and an Ivy
League history professor.

By far the most popular were the talks given by men who knew
how to cook, particularly one young restaurateur from Ohio. "We
even had cooking lectures in the barracks given by John Barbeau,
our little mess sergeant," Frankie wrote in his diary. "He is a swell
guy and you can't help but like him. He's short and stocky with a
very sunny disposition."

POW David Dennis recalled: "You might imagine what happened. After several sessions the attendance at all but the restaurant owner's session dwindled down to a few hard cases. The rest
of us, some two or three hundred, listened raptly to Johnny Barbeau. Funny how I remember those details! He started out well,
telling us some of the funny or interesting anecdotes connected
with his restaurant service, but it soon degenerated into questions
from the floor, such as 'What was on your menu?' or 'How do
you cook corned beef?' The flood gates then opened and everyone, I mean everyone, asked questions like 'How about steak
sandwiches?' or 'Did you ever serve Skippy's peanut butter?' or
'Remember a bowl of Wheaties crammed with heavy cream and
sugar?' Everyone wanted to talk about his dream, because that
was all we could do about our never-ending hunger."

Sonny Fox wrote that in those moments, "recipes began to
sound like poetry."

"We got so used to being hungry," Paul Stern said. "It's hard
to explain [what it's like when you're] starved. You went to bed

hungry, you got up hungry, you were hungry *forever*. Food was all we talked about. We listed all the foods we would eat if we could get back home."

Indeed, some of the POWs literally took pages of their journals to compile such lists.

Pete Frampton remembered the lecture of one soldier with a background as an architect who "helped us build dream castles for our future civilian life."

Frank and Roddie chatted for hours in the gloomy barracks: "Roddie and I have spent some time together discussing expenses for building a small home," Frankie wrote in his diary. "We came to the conclusion that both of us had a sufficient amount of money or would have upon liberation to furnish a very comfortable apartment."

But very soon in their stay at Stalag IXA, a prospective business venture flew around the camp grapevine. "Johnny has big plans for opening a restaurant when he gets home," Frankie wrote one day. "He is the talk of the camp. He and Roddie are going into partnership, and Roddie, being an ex-sign-painter, is drawing up all his plans."

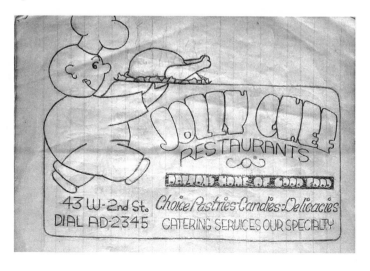

Roddie and the starving POWs dreamed of fine food and one day eating at The Jolly Chef. A desperate man has a chance of staying alive if he has the hope of a future.

Roddie wrote in his journal:

Decided on restaurant business Thursday 15 Feb '45
Definitely decided on restaurant business Friday 16 Feb '45

Planning restaurants like The Jolly Chef gave their lives in captivity purpose. They would resist starvation just as they defied the Nazis. Until liberation came, they would gorge themselves from imaginary menus.

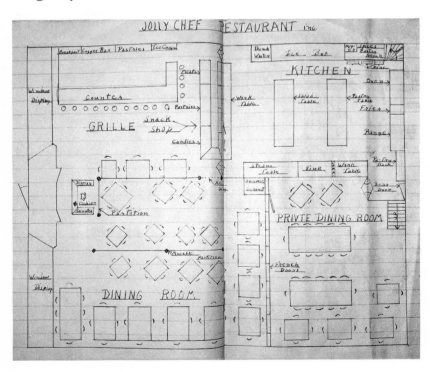

ON A BRIGHT winter morning, the Germans brought in a new group of British soldiers who looked "more dead than alive." By the time they reached Stalag IXA, the British noncoms had been on the road for some sixty days, force-marched from deep in the eastern part of Germany. Some had no boots or shoes, only cloths wrapped around their frozen feet. "Dirt caked their faces and bodies and their clothing was filthier than ours," wrote Richard Peterson.

Word spread throughout the camp that these British prisoners had been marched in savage conditions—many had died along the route, either from illness or exposure, or by being shot.

"Watching the British soldiers' painful progress to the open fields at the rear of the camp," Peterson wrote, "we realized that it could happen to us as well."

This terrifying group of British prisoners brought other news as well. The fate of those GIs segregated at Bad Orb in the "Jewish barracks."

POW Gene Kelch wrote in his diary: "British prisoners arrived from Breslow. Men from Bad Orb forced to work in mines."

Though still a rumor, it was the first indication that something awful had happened to the privates and privates first class back at Bad Orb. What exactly did it mean for those 350 taken from Bad Orb for the special *Arbeitskommando* (work camp) of SS Lieutenant Willy Hack?

Forced to work in mines—the phrase sounded ominous, but Jewish GIs like Lester, Paul, and Skip still had no idea how grim a fate they had so narrowly escaped.

THE DEGREE OF camaraderie and assistance provided to the GIs by their French allies proved invaluable. Having spent so much time behind barbed wire—some more than five years—the

French infantrymen in the adjacent compound were immeasurably better adapted to life under Nazi captivity. Most important to the Americans, the French were regularly receiving parcels from the International Red Cross in Geneva.

"The French looked in good shape, physically, and were willing to give or lend some of their Red Cross parcels," recalled Richard Peterson. "The food contained in these parcels sustained the prisoners of war in the German camps."

The normal Red Cross issue was "one per man, per month." But the American POWs only received one during their time in Ziegenhain. They welcomed the contents of the boxes donated by the French.

"We received our first Red Cross package today," Frankie wrote in his diary on February 8. "Although it was a one-man box, we had to share 4 men to a box. But what a joy. Inside was 5 packages of smokes, a 2 pound can of Oleo margarine, 2 cans of sardines, a box of prunes, box of biscuits, a full can of cocoa, a can of good coffee, a can of powdered milk, sugar, jelly, meat, cheese, etc."

The French, in those early weeks of captivity, also provided another crucial source of hope to the GIs: news from the front lines.

Sonny Fox maintained that his later broadcasting career began with his announcing, in the darkness of his Stalag IXA barracks, military news translated from the French. The French POWs had managed to rig a radio.

"They would listen to the communiqués from the BBC," Sonny wrote, "and then, each day, they would recite them to one of our men, who understood French, at the point where our compound fences touched. He, in turn, would translate the communiqués from both fronts for the five of us who would then be responsible for presenting them each night in our barracks. At six o'clock at night, we would put someone at the door so that if a German guard was coming in I could switch to a lecture on something innocuous. Then I would, by memory, describe what was happening on the Western and the Eastern fronts."

Only through reports on the BBC did the American POWs learn that the Battle of the Bulge had been an enormously costly victory for the Allies, and that even though the men had been captured and were behind barbed wire, they'd played a crucial role in foiling Hitler's counteroffensive to drive to Antwerp.

One particularly grim BBC broadcast had been picked up by French radio, and the news of it reverberated among all the nationalities. On January 27, 1945—the same day that Roddie had risked his life by standing up to Major Siegmann—the Soviet Union had marched through the gates of the largest and most infamous of the Nazi factories of death. The following day the BBC had announced:

The Red Army has liberated the Nazis' biggest concentration camp at Auschwitz in southwestern Poland. According to reports, hundreds of thousands of Polish people, as well as Jews from a number of other European countries, have been held prisoner there in appalling conditions and many have been killed in the gas chambers.

Details of what went on at the camp have been released previously by the Polish Government in exile in London and from prisoners who have escaped. In July 1944 details were revealed of more than four hundred thousand Hungarian Jews who were sent to Poland many of whom ended up in Auschwitz. They were loaded onto trains and taken to the camp where many were put to death in the gas chambers. Before they went, they were told they were being exchanged in Poland for prisoners of war and made to write cheerful letters to relatives at home telling them what was happening. According to the Polish Ministry of Information, the gas chambers are capable of killing six thousand people a day.

Since its establishment in 1940, only a handful of prisoners have escaped to tell of the full horror of the camp. In October last year, a group of Polish prisoners mounted an attack on their German guards. The Germans reportedly machine-gunned the barracks, killing 200 Polish prisoners. The Poles succeeded in killing six of their executioners. . . . When the Red Army arrived at the camp they found only a few thousand prisoners remaining. They had been too sick to leave.

The scale of the Nazi factories of genocide was staggering: historians today estimate that, at minimum, 1.3 million people were deported to Auschwitz between 1940 and 1945; of these, at least 1.1 million were murdered.

Roddie and the other GIs in Stalag IXA didn't yet know the full extent of the Nazi desperation to hold on to their skeletal prisoners. In mid-January 1945, as the Soviet troops and tanks approached the Auschwitz camp complex, the SS began evacuating Auschwitz and its satellite camps. The BBC had announced that the German guards had been given orders to destroy the crematoria and gas chambers, and the prisoners who were able to walk were forced to march to other camps in Germany. More than sixty thousand prisoners marched west toward the city of Wodzislaw Slaski in the western part of Upper Silesia, in Poland, and ultimately toward other concentration camps still functioning in Nazi Germany. Anyone who fell behind, anyone too weak to continue marching, was shot. Prisoners also succumbed to the bitter cold weather, starvation, and exposure on these marches.

More than fifteen thousand died during the death marches from Auschwitz.

WHEN THEY WEREN'T surreptitiously following the news reports of frontline action, and the growing sense of Nazi crimes against humanity, Roddie and the other POWs in Stalag IXA were simply trying to stay sane.

The weeks of malnutrition were changing their bodies, and minds, in unexpected ways. "A physiological fact about starvation is that it reduces the functions of the male glands and enhances those that are more female-like," Sonny wrote, adding that the men's facial hair had turned softer, and "our voices went up in pitch."

A grim joke spread throughout Stalag IXA. "If Rita Hayworth walked into our barracks naked, carrying a corned beef sandwich, there'd be over two hundred guys clawing and kicking and trying to kill each other—just to get a bite of the *sandwich*."

In the pitifully inadequate camp infirmary, men were being admitted daily, suffering from jaundice and full-blown starvation. In early March 1945, the POWs were joined by a medical officer named Captain Stanley E. Morgan. Originally from New Orleans, Captain Morgan had been a POW for six months when he was transferred to Stalag IXA. An Air Corps officer in the 101st Airborne Division, Morgan didn't live in the barracks with the NCOs but was a welcomed addition to the camp leadership.

"Captain Morgan answered medical questions that were on the minds of all of us," Pete Frampton recalled. "Things like, 'How long can we keep going on rations like these?' . . . Six months of this would be about the limit. The poor guy didn't have any medicine to offer us, but he encouraged us in many other ways."*

---

* Many POWs in Stalag IXA estimated their daily caloric intake at close to 500 calories.

Captain Morgan had neither medical instruments nor drugs; initially he didn't even have hospital beds, but he worked out of the office of the American "man of confidence," a sergeant named Elmer Kraske. Born in Detroit, Kraske served with the 179th Infantry, 45th Division; was captured on January 3, 1945, near Wingen in northern France; and transferred to Stalag IXA on January 27, 1945—the same day that Roddie had his confrontation with Major Siegmann.

Serving as the man of confidence was often thankless: the POW liaison would spend his days interacting with the commandant's staff in a fruitless attempt to negotiate better rations, more cigarettes, and improved living conditions. Frankie Cerenzia grew close to Kraske and wrote in his journal that "Roddie also came to be very friendly with him."

Frankie added: "Most times [Kraske] is expected to do things which are literally impossible. Although he tries his utmost to do the impossible, he rarely succeeds, and then he has to put up with all kinds of abuse." Kraske was granted a small office by the Germans and told Frankie he needed an assistant—officially the chief clerk for the American section of Stalag IXA.

"I eagerly accepted chiefly because I craved something to keep my mind occupied," Frankie wrote.

BEYOND THE POWS' dire medical conditions—starvation, frostbite, trench foot, and other diseases—there was also the psychological abyss: some men had given up all hope of liberation.

The look was unmistakable in their hollowed eyes and skeletal cheeks—they were slowly drifting away, somewhere between daydreams, sleep, and death. They'd lost the will to continue even one more day.

Corporal Russell Gunvalson of the 590th Field Artillery Battalion recalled the constant conversations in the barracks about survival: "You're almost ready to give up because of your weakened condition," Gunvalson said. "That's what Germany wanted. . . . The night of that walk from the railroad station to the camp . . . the [German] army probably wishing half of us would have died. Then they wouldn't have to feed us, you know. You don't have to feed a dead man."

Self-harm wasn't openly talked about, but some POWs did privately contemplate ending their lives. "I thought about suicide in my most despairing times, contemplating our open-ended prison sentence," wrote Richard Peterson. "Doubt about the winner of the race between our liberators and the limited capacity of our bodies to tolerate privation tormented me."

Roddie recognized this insidious internal danger. He knew that once a man lost the will to live, he was a goner; once he gave up faith, only slow death from disease and starvation awaited him.

Roddie issued an order that the strongest of his NCOs, what he called the "up" guys, tend to their comrades nearest death, the "down" guys. They helped the weak and hopeless to their feet each morning, got the blood circulating in their freezing legs, forced them to consume the few calories in the black sawdust bread; they urged them to think of home, their parents, sisters, brothers, girlfriends, to visualize the day, fast approaching, when the Nazi criminals would face stern justice and the Sherman tanks would be rolling through the front gates of the stalag.

RECEIVING NEWS FROM the front was so vitally important that Roddie ordered the Americans to build their own radio.

"Roddie, being the communications chief for our HQ Company, knew the value of a radio," Lester recalled. "The radio we had in the POW camp was put together by the radiomen who had been trained by the army for that job—from parts smuggled into the camp by the volunteers who went to work in the town of

Ziegenhain, ostensibly as a gesture to the German people, but actually planned to obtain information and any materials that could be brought back to camp."

NCOs in Stalag IXA were not obliged to work for the Germans, but volunteer work parties were organized daily—under a parole system. Sometimes the Germans offered an incentive, as Frankie wrote, "of ten cigarettes, an extra ⅙ of slice of bread, or ration of soup."

Lester recalled Roddie's thinking: "We needed a radio, but it would be too large to hide in the clothing of these men. So the component parts had to be small enough to be concealed, quickly, so that they could be assembled back in camp and hidden during the barrack inspections by the guards."

The radio surreptitiously assembled in Roddie's barracks allowed for more direct monitoring of the approaching Allied armies.

"Feb. 23, 1945, was a big day for us," Frankie wrote in his diary. "We had been patiently looking forward to the Allies starting their big drive in the West to cross the Rhine and begin the final phase of the war. Today it happened. The big drive had finally started and was meeting with success. Now everyone was in high spirits again."

From the end of February, the men regularly gathered around the makeshift radio to track updates about General Patton and his rapid advance through Germany. Itching to go on the attack, Patton defied General Eisenhower's strategy of "aggressive defense" and kept "moving towards the Rhine with a low profile." He told his staff that the Third Army was going to carry out an "armored reconnaissance," using seven divisions.

Roddie and the other men listened as Patton's Third Army tore through the same terrain where the 106th Division had fought, crossing the Schnee Eifel's three major rivers, which were "swollen by the snow and rains of the worst winter in thirty-eight years," before capturing the cities of Prüm and Bitburg by March 1. They captured Trier a day later.

Eisenhower and his staff at SHAEF had estimated it would take *four* full divisions to break the heavy Nazi resistance and capture Trier, but Patton sent Ike a sarcastic message: "Have taken Trier with two divisions. Do you want me to give it back?"

At last, Patton received the command from General Omar Bradley he'd long been waiting for. He was given full authority to move at a lightning pace and "take the Rhine on the run."

On March 7, the intact Ludendorff Bridge at Remagen was captured by the First Army, and on March 10, Patton's Fourth Armored Division reached the Rhine River, just to the north of Koblenz. Patton's forces had advanced a stunning fifty-five miles in less than two days. By March 22, Patton had eight divisions massed along the Rhine from Koblenz to Ludwigshafen.

Patton's campaign west of the Rhine was over. The general made certain that Third Army engineers were ready with pontoon bridges so his forces could cross the river before Montgomery, and early on March 23, "six battalions were over the river [with] a loss of only 28 men killed and wounded, while other infantry and engineer units had crossed just to the north, at Nierstein, without opposition." After heavy Luftwaffe raids on the Third Army pontoon bridges during the day, Patton called Bradley. "For God's sake, tell the world we're across," he said. "I want the world to know Third Army made it before Monty." At Bradley's headquarters that morning, the announcement was made that the Third Army had crossed the Rhine at 10 p.m. on March 22, "without benefit of aerial bombing, ground smoke, artillery preparation, and airborne assistance."

On the day his first troops crossed the Rhine, Patton issued his now-famous General Order Number 70 to the Third Army and its supporting XIX Tactical Air Command:

In the period from January 29 to March 22, 1945, you have wrested 6,484 square miles of territory from the enemy. You have taken 3,072 cities, towns, and villages, including among the former: Trier, [K]oblenz, Bingen, Worms, Mainz, Kaiserslautern, and Ludwigshafen. You have captured 140,112 enemy soldiers and have killed or wounded an additional 99,000, thereby eliminating practically all of the German 7th and 1st Armies. History records no greater achievement in so limited a time. . . . The world rings with your praises; better still General Marshall, General Eisenhower, and General Bradley have all personally commended you. The highest honor I have ever attained is that of having my name coupled with yours in these great events.

The following day, at Oppenheim, Patton crossed the Rhine on a pontoon bridge. Halfway across, looking down into the mighty river, he opened his trousers "to take a piss in the Rhine. I have been looking forward to this for a long time," he wrote in his diary, "then I picked up some dirt on the far side in emulation of William the Conqueror."

Patton's act of contempt was done in full view of his men and news cameras.

"I have just pissed into the Rhine River," he wrote Eisenhower. "For God's sake send some gasoline."

Roddie and the other POWs celebrated Patton's seemingly unstoppable momentum.

"We've learned that Gen. George Patton has captured 70,000 Germans, Cologne has fallen, the Rhine has been crossed in a number of places, and in 3 very important places that we know of, Cologne, [K]oblenz, and Dusseldorf," Frankie wrote in his diary. "At night and all day we can very clearly hear the battle sounds of artillery. Our boys are not over 85 miles away from here. Every single day we can see or hear thousands of our planes flying into the very heart of Germany and laying many noisy eggs. By April 5th I expect to be eating good army chow till I burst, and by the middle of May I expect the most hated nation in the world to be crushed completely. I hope and pray I'm right."

TWENTY-TWO

D AY BY DAY the Allied bombing got closer. It gave all the
POWs hope. They could hear the sounds of artillery and
tanks, and the Allied bombers—the RAF by night and the US
Army Air Forces by day.

Sonny Fox wrote that as the front grew closer, "a high point" of
the men's day became watching the contrails of the B-17s as they
made their way eastward over Germany on their daily bombing
runs. "Sometimes," he wrote, "there would be hundreds of them
and the deep, pulsating noise of their four engines was the affir-
mation that the war was moving into its noisy final stages.

"As the front moved closer to us, our fighter planes were buzz-
ing around, strafing German columns, freight trains and pretty
much anything that was moving. We crowded the entrance to the
barracks to watch as much of that show as we could. Though we
restrained ourselves from cheering so as not to piss off our captors, there is no doubt that the sight of this armada of American
power was a tremendous boost to our morale."

Still, one of the ways they liked to harass the guards was to sing a song of their own creation—to the tune of the "Battle Hymn of the Republic":

We're a bunch of Yankee soldiers living deep in Germany.
We eat black bread, a little soup, and a beverage they call tea.
And we have got to stay right here till Patton sets us free
And we go rolling home—
Come and get us, Georgie Patton.
Come and get us, Georgie Patton.
Come and get us, Georgie Patton.
And we go rolling home.

The song would enrage the guards and result in "screaming and rifle butts"—the "Germans like to scream a lot," John Morse remembered.

By the middle of March, the sounds of the Allied air attacks grew so close that the entire camp shook with reverberations. "For the past 24 hours there has been terrific bombardments very close to us," Frankie wrote in his diary on March 19. "It hasn't stopped once. I'm inclined to think that something big is up. I sure hope I'm right."

On March 20, Frankie wrote: "Boy, am I pooped out today. I went out on that wood detail outside of camp today and it's taken everything out of me. I never realized how weak I really am until today. Nevertheless, going out there in those fresh green woods was really wonderful. Gosh how I wish I were free. Especially when I was out there sitting under a tree during a rest period. I got to thinking of Lucy, picnics etc. I'd better just keep on praying I guess. That terrific booming is still going on out there but so far nothing big has come up. As usual, the Lord was kind to me again today and I'm fairly contented. Let tomorrow be still better oh Lord, please. Good night Lucy, my darling."

His next entry reads: "March 21, the first day of spring, was warm and sunny and most of the men were outside in the sunshine. More Allied planes appeared in the distance. The men realized that the front lines must be creeping very close indeed."

In the middle of the afternoon, they saw a squadron of US Army Air Corps P-47 Thunderbolts—the fighter designed to accompany Allied bombers deep into Europe and return to England. On the way home, Pete Frampton remembered, the Thunderbolt pilots were "given clearance to attack any targets of opportunity, and these boys took a hearty interest in what looked to them like a first-class, large size German Army Barracks Complex, which is exactly what our prison was, once upon a time."

On the first pass, the GIs out in the compound stood waving hello to the pilots.

But then one of the German guards decided to shoot at the P-47 with his machine gun.

The Thunderbolt pilot swooped down suddenly, strafing the camp with his six wing-mounted .50-caliber guns. The bullets tore through both the French and American compounds, sending men scattering, diving for cover.

"When they started to shoot, we were stunned, and then we hit the deck, prone, flat, small as possible," said Pete Frampton.

"Every one of us hit the ground solid, and boy, I thought this was our finish," Frankie wrote. "Just as the bullets hit our barracks, he stopped firing."

"We were lucky," Sonny recalled. "We were digging slugs out of the wooden bunks for days."

The damage in the French compound was far worse. Eleven French POWs were killed and forty-nine injured.

"As we watched the bodies being borne past our compound, the sadness was accentuated by the realization that after five years as POWs, they had been killed by their ally as liberation was at hand," Sonny wrote.

The P-47 strafing left a dark pall over the camp—the men were in a state of shock and high anxiety. The front was drawing nearer every hour, and they could easily be subject to continued Allied bombing and strafing.

Liberation would eventually come, they knew. But had they come all this way, survived the battle in the Ardennes, the forced marches, the Christmas Eve bombing, the starvation of the camps, only to die by the friendly fire of their own air force planes?

The men knew it was only a matter of weeks—perhaps days— until Patton's armored divisions reached them . . . but would they survive long enough to see freedom?

"It's very apparent that our front lines are getting extremely close or those planes wouldn't be concentrating on this one area," Frankie wrote in his diary, adding ominously: "Our lives are in danger here every single day, and some of us may never reach home at all."

THE BOMBING NEAR the camp and the increased risk of another strafing led to some of the prisoners demanding action. "We saw flashes of artillery or bombs landing in the night sky," wrote John Morse. "Right after the P-47 incident we requested the commandant let us paint POW on the barracks roofs. He declined. We thought more air attacks would come as the Stalag could be mistaken for German barracks."

On March 24, Frankie wrote in his diary: "In the afternoon the fighter planes were over again. They did some strafing and then left. We got our bread and margarine and just at that time, a slew of bombers came over and really gave us a scare because we thought they'd drop their load on us. But they didn't. They did considerable damage to the little town just outside of the camp."

There were also rumors swirling that before the Allied liberators reached the camp, the Germans would force-march all the prisoners—now numbering approximately sixteen thousand—deeper behind the front line into Germany.

John Morse was in the same barracks as Staff Sergeant Sam Harris, the young artillery gunner from the 333rd Field Artillery Battalion. A group within the barracks held a meeting, Morse recalled, "and decided we had to break someone out to warn our people that this was a POW enclosure." They needed to tell the Allies to "stop strafing us and come get us. Three guys volunteered to try a jailbreak. Two white guys and our one black guy. We tried to time the appearance of the guards on their rounds. Two rows of barbed wire fence with guard towers every so far. It wouldn't be easy. To tell you the truth I was too damn chicken to try. I did go out with our team to see them off, hiding behind a building late at night. In spite of the cold I could see the sweat on their faces. The black guy moved first with a final: 'Let's go!'"

There is some disagreement about the exact details of what transpired during the escape in the early hours of March 29; the darkness of night, adrenaline, and fog of memory leave certain facts unclear to this day. Indisputably, the night turned fatal. But even a thorough US Army homicide investigation conducted in October 1945 gathered conflicting eyewitness testimony.

Staff Sergeant Kevin Lee Finneran of the 423rd Regiment was interviewed by the War Crimes Office of the War Department. In the investigation, he told his version of the night: "After dark . . . Sam Harris, who was quartered in the same barracks that I was at Stalag IXA, attempted to escape in the company of one or two other prisoners of war. We had heard that the American troops were at Treysa, which was only 17 miles away."

Finneran, Harris, and another soldier, Staff Sergeant Peter Scandalis, planned to try to escape to Treysa that night to contact

US forces and alert them to the presence of thousands of POWs in Ziegenhain.

"Our barracks was about fifteen yards from a latrine which was right up against the outside barbed wire fence surrounding our compound. This fence was made out of four-inch squares of barbed wire. Between that fence and the outer barbed wire fence made out of the same sort of squares was a space of about fifty to sixty yards wide, which was used in the daytime for recreational purposes. Our plan was to go into the latrine which was situated right up against the inside fence and get out the window so that we would not have to climb the barbed wire."

At approximately one o'clock in the morning, the escapees exited their barracks through a window, then hid in the camp latrine. They broke the barbed wire that enclosed the small window. According to Finneran, heavy clouds had been hiding the moon, and it was very dark in the camp compound. Sam Harris climbed through the window first and jumped out onto the ground.

"But just as he jumped the moon came out from behind the clouds and lighted up the scene as though someone had turned on a light," Finneran said. "The moment he landed, the guard, who had been stationed near the gate in the fence about fifteen yards away, began to yell and then several other guards started yelling. We heard three shots at him after that, and Scandalis and myself ran out of the latrine back to our barracks and mixed with the other men in the barracks."

Sergeant Christian Wicks, a POW from Napa, California, gave investigators a grim but distinctly different eyewitness account: "Sam let himself down into the pit under the latrine and had crawled down to the end of the pit, escaping through a hole there that was used when they pumped the pit out. From the hole he crawled down a ditch that led away from the compound and toward the outside fence. He had gotten about 75 yards down the ditch when a guard, who was patrolling the area to the north of the compound, spotted him. The guard fired one shot at Harris, and apparently struck him in the shoulder, for Harris got up in sort of a stooping position. The guard then ran over to where Harris was—by this time he had fallen to the ground again—and put his foot on Harris' back, then fired another shot into his head. The guard was about 100 yards away at first shot. It was about 11 o'clock at night. However, the moon was very bright, and we were able to see what happened."

Frank Smith, a tech sergeant from Pittsburgh, testified he was another of the escapees and spoke with Sam in the middle of the attempted breakout: "An American colored soldier by the name of Sam Harris and myself attempted to escape from the German Prisoner of War compound at Ziegenhain, Stalag IXA, by crawling over the compound fence. Just after crossing the fence, I heard German guards being called. I told Sam that we had better get out quick. Sam seemed to become frightened and said he was going to stay there between the fences in a bomb shelter."

Smith then left Harris and crawled back on his hands and knees about twenty-five feet to a ditch. "A German guard came up to where Sam Harris was hiding and ordered him out. Without trying to run or attempting to effect further escape, Sam gave up to the German guard. Instead of taking him back to the compound, the guard shot him. Sam hollered after he was shot, and the German guard shot him again, killing him. About two hours after Sam was shot, a German officer found me. After kicking me several times, he led me to the Orderly Room."

Frank Smith's sworn testimony is clear in his assertion that Sam Harris had surrendered and was executed in cold blood. Still others say that Sam was callously left to bleed to death in a muddy trench.

One POW said he saw Harris successfully scaling the fence en-

closing the American compound and was in the process of climb-
ing the outside barbed fence when "an alarm went off, a watch
tower spotlight was set on the scene of the escape, and the guard
in the watch tower shot Harris with a rifle. . . . Immediately there-
after the guard who had done the shooting went up to where Harris was lying, prodded him with his foot, then kicked him into a
nearby air raid trench where he remained all night."

It's perhaps understandable, given the darkness in the camp and
the chaos of the moment, that there are so many disparate memories of the shooting's details; yet almost every single POW interviewed in the judge advocate general's investigation was clear
that the killing of Harris constituted a war crime.

Staff Sergeant James Lindow of the 422nd Regiment said that
while trying to escape, "Harris was shot, kicked into a ditch, and
bled to death during the night." And Sergeant James M. Hen-
nessey, in a list of atrocities committed by the Germans and personally witnessed by him, wrote: "A Negro Staff Sergeant, Sam
Harris, was shot and left laying in a trench until he bled to death
at Ziegenhain, Stalag IXA."

US Army investigators pursued all information about the guard
in an attempt to locate him. Several POWs said that the German
was a sergeant and one of the regular camp guards—they only
knew him by the nickname "Hook Nose."

It was clear to the POWs who were eyewitnesses or had firsthand knowledge of the escape that the shooting of Sam Harris
had been a case of murder by one or more German guards.

The man of confidence, Staff Sergeant Elmer Kraske, was not
an eyewitness to the shooting but was shown the body hours
later. He was told that Harris was shot attempting to escape. "I
saw bullet holes in the top of each shoulder coming out his back
near his lower lungs," Kraske said. "I do not know whether he
bled to death or was instantly killed. An English medical officer, a

captain, told me at the time that he concluded from his examination of the body that Harris had been murdered."

The murder of Sam Harris had a chilling impact on the entire camp, leaving many men frightened, angry, and determined to exact justice.

Richard Peterson, in his memoir, wrote of the stark ramifications for the whole camp: "The fury of the guards resulted in the longest and most thorough count I ever experienced. The guards forced us out of the barracks into an empty field in the usual five-man column. They counted us repeatedly. . . . We were there for three hours. It was so cold I thought I would never get warm again. Apparently, one man attempting to escape had managed to get through the first fence and into the dead zone, only to be killed. The guards carried his frozen body past us for all to see."

According to some, the Germans threatened reprisals if any other escape attempts occurred. "Two guards carried our dead hero's body back and forth in front of us while the camp commandant lectured us on the futility of trying to escape," wrote John Morse. "One more attempt and six men from the guilty party's barracks would be executed."

TWENTY-THREE

A S THE EXPECTATION of liberation grew greater, so did the apprehension and fear. The POWs sang daily for "Georgie Patton" to come and get them, but would they even live to see the general's tanks and jeeps barreling through the camp gates?

The proximity of the front lines heightened the anxiety of the Germans, and that feeling was palpable to the POWs too.

"We were seeing P-38 pursuit planes diving at Ziegenhain at 11:00 a.m. every day," Roddie wrote. "The little cub planes which normally spot for field artillery were coming over occasionally. The Germans were downright nervous and edgy."

By then, the starvation rations had caused most men to drop 50 to 80 pounds. It wasn't clear how many more days they could survive under such conditions.

"We got word that Patton is driving like mad," Frankie wrote in his diary. "Things are finally beginning to look up and I hope and pray we're free men soon. This is the most horrible experience I've ever had in my life. 5 men died today in the last 24 hours from starvation and that doesn't make me feel any better either."

There was great whispering that something big was afoot. But none of the men knew exactly what. Frankie's diary entries convey the fear bred from uncertainty:

———————————

MARCH 26. *Today is a tense day, is all I can say about it. Something is up and none of us know what it is. Frankly, I'm worried. Some Germans are already leaving the camp. I'm not going to say any more about today, tomorrow I should learn more.*

MARCH 27. *Another tense and quiet day but oh, what news. Our armies are only 35 kilometers away and swiftly moving closer. That puts them about 22 miles away. Something is happening all right and unless they should move us out, we'll be free by Easter. That's what I'm so darn worried about. They can move us and if they do, we're all sunk until the war's over. God, I wish I knew what all the tension and whispering is about. Perhaps tomorrow?*

MARCH 28. *What a day. This morning, I received the good news that the allies are only about 15 miles away. Then what I'm most fearing has happened. We were told we would be moved out tomorrow. Somehow, I knew that was coming but I won't believe yet that my prayers would fail me.*

———————————

Medical officer Captain Stanley Morgan confirmed what Roddie and the other camp leaders suspected: any march would prove fatal to the men. But Captain Morgan believed that the very ill—those near death—would be left behind.

"It was important that we not leave the POW camp because

once outside you were subject to being shot by your own planes, which were everywhere, and Germans and civilians," Roddie wrote.

Of course, Roddie was aware that the German plan to evacuate the camp was against the Geneva Convention's stipulation forbidding placing prisoners in danger. He also was aware—thanks to the BBC broadcasts—that as the end of the Third Reich was drawing near, the Nazis were deliberately force-marching tens of thousands of men—from prisoner-of-war camps, work camps, concentration camps, and death camps like Auschwitz—deeper and deeper into the heart of Germany.

"The Great March, as we called it, was to begin at sunrise the next morning," Richard Peterson recalled. "We seldom spoke of death in the camp. Perhaps none thought we would die there because we had escaped death in combat. This changed when we were ordered out on the roads. Then, I thought we would all die."

Frankie's diary includes a dark notation: "I wrote a letter to Lucy for Louie to take back. It sounds more or less like a last will and testament. Who knows? Maybe it is."

He wasn't the only POW to think he was in his final days. Peterson wrote what he felt was likely "the last letter to my family, made a short will and tried to make my peace with God."

One more time, Roddie refused to give in to defeat, not when liberation was so near. There was no way he would allow the men under his command to comply with the Nazi order. "We're just too weak to go on a long march," he told Lester.

Roddie came up with a strategy—much as he had with the order to deliver up the Jewish noncoms. And, just like last time, this plan carried considerable risk.

"No one marches out of the camp," he said. "No one."

When the order came to fall out, Roddie told the American POWs, they would delay, hide, pretend to be too ill to march—

anything to stay in the camp. Every third man was ordered to feign illness and fall to the ground. The men to the left and right of him would help him back into the barracks. Perhaps by delaying the count and stalling, by dropping to the muddy ground in agony, they could prevent the German attempt to evacuate the camp.

"The danger was in the confusion," Peterson wrote. "If the guards became inflamed enough, some men could be killed, as happened in our early days of captivity. The risk to some was worth the benefit for all."

The fear was almost universal among the men. Few could sleep as the morning roll call approached.

At 0600 roll call, as Roddie had ordered, no one moved from his bunk.

"*Raus! Raus!*" the Germans screamed.

At 0615, the men were still in the barracks, trying to delay as long as possible. As they were milling around, guards burst in with snarling German shepherds. The dogs snapped at the POWs, and the Germans shouted in anger.

"We moved out in a rush trying to avoid the sharp teeth of the big animals," Peterson recalled.

Roddie had ordered that as many men as possible should hide. Some hid underneath the barracks; some hid in the latrines. This frustrated the Germans in their five-man count.

As the Americans formed up slowly into ranks, they saw that the French were marching out of the camp.

When the order came to march, the Americans moved only a few paces before GIs started to collapse in apparent pain. "I joined them, groaning, and fell into the mud about a hundred yards beyond the barracks," Peterson said. "The guards moved up and down the ranks screaming and kicking those on the ground. The dogs growled and barked. Some of the men were being mauled. We moaned louder."

But Roddie and the rest of the men continued to carry the supposedly sick men into the barracks. And in no time every American was back inside. "We spent about half the day playing sick," Frankie wrote, "and all this time Patton was drawing closer and closer."

But *playing* sick wasn't enough. As the Germans grew more impatient, Roddie ordered the POWs to make themselves ill.

"After we came out of hiding and were assembled, I told all our people to eat grass and dirt and get sick," Roddie wrote. "I told them not to fake it but to really get sick. They got so sick one of the German guards cried. Half of our people were sick and the other half carried them back into the barracks."

Some men swallowed soap to foam at the mouth. Captain Morgan even prepared a concoction from supplies in the infirmary that made several of the soldiers vomit.

Many were in convulsions while still others rolled on the ground, moaning and muttering incoherently.

Back and forth the men continued, carrying one another into the barracks. All while the dogs kept snapping and mauling at them.

After the French had left their compound, the Americans watched the Germans march out the British soldiers. Then the Russians. Then the Serbs.

Every time the GIs formed up, Roddie would give the order, "Break ranks!" and the men would fall to the ground or dash back to the barracks.

By now the Germans were getting furious at the Americans' stalling. Some started firing their rifles into the air.

They ordered the Americans to send at least twenty men from each barracks out immediately.

After several hours, the exasperated Wehrmacht commandant, Oberst Mangelsdorf, stood face-to-face with Roddie.

The Germans realized they had only a brief window of time before their own escape route out of Ziegenhain would be blocked. Patton's army was too close. Sherman tanks and half-tracks were closing down the roads.

Roddie wrote: "Finally the old German colonel came down and threw up both hands and said we had won and could have the camp. He was leaving. The Americans were the only group that were not marched out."

Paul Stern recalled: "[Roddie's] sense of duty, responsibility, and devotion to the soldiers under his command went far beyond his own personal safety. All the American prisoners at the camp were saved due to his outstanding courage."

The Wehrmacht officers and guards left the camp. But there was still an imminent threat.

Roddie ordered that all the men go back into the barracks.

John Morse wrote: "We were told, 'Stay away from windows, use the sinks for latrines, no lights, no fires, no smoke or cigarettes and be quiet. If the German troops who might pass this way ever look in here and see what we have done, we are all goners.' We did as we were told. We hid all the rest of the day."

But suddenly a few Waffen-SS men appeared. Roddie and the Americans remained out of sight, fearful of these ruthless Nazi fanatics.

More and more SS men arrived in columns and then in armored vehicles. The GIs were terrified that the Waffen-SS might massacre them as they had the unarmed American prisoners at Malmedy. After more than one hundred days in captivity, the POWs knew the camp intimately—every corner and crevice.

Roddie and the other noncoms—nearly 1,300 men—managed to remain silently hidden from the Waffen-SS. Roddie's diary entry for the day contains just a single word: "Hiding."

BELIEVING STALAG IXA to be deserted, the SS troops hurriedly fled eastward, farther into the heart of the Reich.

Within hours, Frankie wrote in his diary, "We found that we owned the camp. All the Germans had skipped out. We posted our own guards all around camp, so men wouldn't get out of control and start looting. We continued to sit back, waiting and praying. . . . The camp was quiet and tense. There was a tension in the air as thick as butter. Tension was so great that hunger was forgotten. Lord was I praying."

Then Frankie continued: "On the 30th of March, Good Friday, things started popping. GI Joe was expected to arrive at any minute. I was given an armband and told to stand watch."

With every passing hour, the POWs could hear sounds of heavy American artillery getting nearer and nearer, booming 155 mm cannons. On that Friday morning the men saw a small scouting plane used as a forward spotter. They soon heard sounds of rolling armor, half-tracks, and tanks.

The houses of Ziegenhain visible to the POWs were now flying white flags of surrender.

At 1530 on March 30, 1945, tanks rolled down the road from Ziegenhain, then turned at the guard tower and into the camp.

In his diary, Frankie called that moment, "the greatest thrill of my life. I cried with joy, for there, rolling into the gate was a grand American jeep. Chaos broke loose in the camp. All prisoners busted out of their compounds and greeted Johnny Doughboy with hugs and kisses and genuine tears of joy."

One Sherman tank burst right through the camp fence—they

didn't wait for the gate to be opened. It was indeed Patton's Third Army—the Sixth Armored Division. To the ragged, starving POWs, Patton's men, in their clean infantry uniforms, "looked like giants."

Paul Stern recalled that as the American armor rolled in, "the boys ran to the fence, kissing the tanks."

The POWs were singing their "Come and get us, Georgie Patton" song, and they scrambled onto the Shermans and half-tracks, yelling and cheering.

Sonny Fox stared at the first GI liberators in awe. "Man, are they *fat*," he said. At his starvation weight of 104 pounds, he was used to seeing nothing but skeletal GIs in sagging, filthy olive-drab uniforms.

The men of Patton's army were shocked at the sorry condition of the men. Until proper food supplies could arrive, they distributed boxes of their personal K rations, which the POWs devoured.

Some of the German guards who'd escaped were recaptured by the US forces on the roads out of the camp and brought back to Stalag IXA. At least one German guard—"a real SOB"—received vigilante justice from the POWs.

"We were mad—when we got liberated, we were mad," recalled Russell Gunvalson. "We found the warehouse full of Red Cross packages meant for us and we didn't get them." The men also discovered a trove of their own letters home that had never been sent. "That made us really, really mad. And some of those guys . . . they wanted to get even with somebody."

By several accounts, the Wehrmacht guard known as "Hook Nose," who'd shot Sam Harris to death, was beaten to death by angry GIs.

But most of the men were in too much of a joyous mood to look for retaliation.

In addition to being Good Friday, March 30 was also the second day of Passover.

Paul Stern reminded Skip Friedman of the vow they'd made that they'd survive this ordeal and be celebrating the Passover holiday in freedom.

"An American Sherman tank had pulled up to the gate at Ziegenhain," Skip recalled. "Les and Paul and I went out to greet it. An American tanker sticks his head out of that tank, and he yells, 'Anybody here from Ohio?' I shouted back, 'Me!' And Paul says, 'You have any food?' And the captain in that tank pulls out some hard crackers and throws them to us. Paul looks at it and says, 'I told you we'd have matzoh for Passover.'"

TWENTY-FOUR

W ITH LIBERATION, THE MEN had fully restored order to the camp and proceeded to set up a proper military organization again with duty assignments and a rigorous chain-of-command.

"My stomach is full again and I'm too happy to say any more," Frankie wrote in one diary entry. "I have an awful lot to talk about when I return to Lucy." The men waited for days, milling about, wondering when they'd be taken out of the camp. Stalag IXA was the first major POW camp liberated by the Allies, and no one at SHAEF was prepared with any sort of evacuation plan. "The result was that although we were no longer prisoners, we were still in the prison camp, and, as it turned out, would be for another week. The diet improved somewhat, though more by quantity than quality," Sonny wrote.

As the US forces advanced, they captured large numbers of retreating German troops and brought them into Stalag IXA, where the former POWs found that the tables had suddenly turned. Toting M1 Garands, GIs herded the bedraggled Germans into formations. When some American soldiers came through asking for volunteers to guard the increasingly large number of German prisoners pouring into the open areas inside the prison gates, Sonny jumped at the chance. But he wasn't the least bit eager to "torment my tormentors." His sole motivation was the prospect of getting better rations.

Sonny took up a position in the guard tower—which the POWs had for so long feared and derided as "goon boxes." "The very machine guns that had been trained on us [were] now trained on the thousands of Germans milling about below," Sonny recalled. "Since we were quite depleted from our starvation rations of the past several months, we did not make very good guards." Sonny admitted to constantly falling fast asleep at his post in the guard tower.

The GIs had been in their same uniforms for five months—some, like Roddie and Frankie, had tried to wash their clothes regularly, but the POWs were still in need of fumigation, which came in the form of a strong disinfectant sprayed from huge firefighters' hoses.

Even before proper US Army chow trucks arrived, the GIs discovered that their Wehrmacht captors had hidden "huge mounds of potatoes stored in straw"—and they were fed "their first real meal since before we were captured." The trouble was their digestive systems could barely tolerate even boiled potatoes. Almost every man was stricken with diarrhea, headaches, and disorientation.

"We had never seen a potato in our daily slime," John Morse said. "Food—real food—was too much for our systems." Morse soon found he could no longer walk, or even stand up. "I had gone from 170 pounds to 90 pounds as a guest of Adolf."

Skip Friedman and Paul Stern made a bold decision—a risky one too, considering their badly emaciated and weakened condition. "When American tanks busted into the camp," Skip recalled, "the war was still going on around and we realized it could be a long time before we were finally behind American lines." Skip, Paul, and two other noncoms left on a mission to get some better food, clothing, cigarettes, and medicine.

Skip and Paul learned of a major army depot in Bad Homburg, near Frankfurt, 90 miles away, borrowed a truck, and barreled down the highway to Bad Homburg without stopping. It was a point of pride for Skip and Paul that despite their weakness, after over a hundred days in captivity, "we were tough enough to be able to do it—we were among the very few guys strong enough to do it—and brought back a lot of stuff to camp," Skip recalled. "Candy, food, rations, some clothing, cough medicine, whatever we could get our hands on." There weren't many POWs who had not only the physical strength but also the motivation and mental clarity to make that 180-mile supply run.

Most of the POWs were in a post-liberation daze. Waiting for medical care—almost everyone suffered from the "trots"— dysentery was rampant, and in some cases severe. The combat medics among the POWs, like Paul, knew that some of these cases could quickly cause death. The Transportation Corps sent a couple of two-ton trucks into Stalag IXA to carry out guys who were in the worst shape. Early Sunday, April 1—Easter morning—the Medical Corps had arrived with ambulances and medicine, and a group of the most severely ill POWs, including John Morse and Pete Frampton, were trucked to an airfield near Giessen.

"We arrived at about the same time as a group of about six C-47 transports began to land," Frampton said. "The planes killed their port engine and let the starboard engine idle while we jumped aboard. In less than ten minutes, we were airborne and heading back to the west coast of France." They were returning to where their European tour of duty had begun: Le Havre, on the Normandy coast.

The men who were left waiting in the camp—including Roddie, Frankie, Lester, Paul, and Skip—tried to find moments of humor in their grim situation. Lester, at six feet, had dropped from 180 pounds to 120. With a sardonic smile, he realized that he now weighed about the same as one of his petite girlfriends back home. When they stripped off their shirts to shower, they couldn't hide the shock at their own skeletal physiques. Someone joked they looked like walking xylophones or piano keyboards—"You could have played any kind of music you like on our ribs," said Russell Gunvalson.

In the days after Patton's tanks burst through the camp gates, the GIs had time to reflect on their final moments in captivity. Lester was still marveling at Roddie's selflessness and courage standing up to Siegmann and refusing under threat of death to obey Sieber's order to evacuate the camp on a final death march.

"We weren't *liberated*," Lester said. "We escaped." Years later, Lester would often describe it as one of the "greatest mass escapes in World War II"—more than 1,200 infantrymen liberating themselves from Nazi captivity.

Roddie wanted neither credit nor accolades—he simply hoped to enjoy the peace. He spoke little about his feelings now that he was a free man—not even to his closest buddies, Frankie and Lester. He preferred to let his pencil do the talking. He wrote succinctly in script in the pages of his diary, decrying the terrors of war and thinking of friends from whom he'd been separated, like Sergeant Jack Sherman and Private Ernie Kinoy—while still in the grim barracks of post-liberation Stalag IXA.

"I have made new friends and lost some," Roddie wrote. "I don't know whether all of my boys are alive or not. But I pray that they are. It all seems sort of a bad dream—a *very* bad one. I have been overseas a little less than six months."

By now Roddie seemed to see his role in the war—the bloodiest in human history—in philosophical, as well as spiritual, terms.

"I'm just a little guy, but war isn't right," he wrote. "Lives upon lives are lost, people forget God more and more. It seems as if someone should get wise and make the whole world a Christian world. Let God be our Commander and we should all live our lives as he lives his. It's bad and it's got to be changed soon."

AFTER SEVEN DAYS as free men still living behind barbed wire, a collective cheer rang out when the US Army Transportation Corps sent a long row of empty trucks through the gates of the stalag. To Richard Peterson's eyes, those "utilitarian trucks" were a thing of great beauty. "Foot soldiers have a one-sided love affair with them," he wrote. "These ungainly, generally dusty, dull khaki-colored, uncomfortable and graceless vehicles are adored by the walking soldier. Trucks always arrive in a cloud of dust, making a great noise from the chatter of wheels and whatever loose attachments they carry. There is the clatter of tarpaulin bows, shifting bangs of folding benches and a general rattling from other gear. To an infantryman, trucks not only promise an end to the day's march, but a journey to somewhere better. In almost three years in the infantry, I never remember wanting to be marching or enjoying the scenery or knew where I was. The trucks were the magic carpet to somewhere, anywhere else."

It took until April 10, 1945, for most of the men to be taken out of Stalag IXA. Down the desolate German roads the trucks roared, the GIs taking in the white flags of surrender fluttering from open windows amid the largely ruined German towns and villages.

Roddie, Frankie, and Lester rode side by side, miserably cold in the open trucks. Frankie pointed out, with some pride, the damage "the American army was doing to get this war over with fast. All German towns and villages were literally smashed. Nothing but a mass of debris. White flags of surrender hanging from every window of every house we passed." And in every town, they saw the young "frauleins dressed and painted in their prettiest to impress and attract GIs."

Richard Peterson felt "almost drunk with happiness" that the months of torment had ended at last. "The promise of clean clothing and army food, which I had never thought of before as luxurious, was almost overwhelming."

Frankie remarked to Roddie and Lester that he was astonished by the Nazi determination to continue in this losing cause. "With all the food supplies that the US Army uncovered, it's pretty clear that Germany is prepared to fight to the bitter end," he said.

After hours in the rattling uncovered trucks, they reached the "remains" of a Nazi airport where many C-47 transport planes were waiting. Skip Friedman—always an avid amateur historian—knew that Giessen had been one of the biggest Nazi air force bases in Western Europe. "In Giessen we got into C-47s—my first airplane ride ever," he said. Many of the men—including Roddie—had never been airborne before.

The flight to France took a little over two and a half hours in those heavy prop planes—but Frankie called it "the best trip of any kind that I ever made in my life so far." From the landing strip in Le Havre, they piled back into trucks and were driven to a RAMP camp—quickly learning that the acronym stood for "Returning Army Military Personnel"—which was set up with many thousands of tents. These were temporary US Army tent cities located principally around the French ports of Le Havre and Marseille, set up in the days after the Normandy invasion in June 1944.

"Out of Germany at last!" Frankie jotted in his journal as they arrived in the camp. This RAMP was called, with no little irony, Lucky Strike. The men learned the names of some of the other RAMPs and quickly dubbed them "cigarette camps," having a good laugh together over Camp Chesterfield and Camp Philip Morris and Camp Old Gold—a punch line only weary POWs could fully appreciate, tobacco having been such a precious commodity within the stalags, especially for lifelong smokers like Frankie. The US Army had chosen brands of cigarettes primarily for security: referring to these RAMPs without any indication of their geographical location in Normandy or on the French Riviera was a safeguard, as any Nazis eavesdropping or monitoring Allied radio traffic would think that cigarettes were being discussed.

At Camp Lucky Strike, all the former POWs were registered and debriefed. A field hospital staffed with doctors and nurses checked all the men for lice, malnutrition, respiratory ailments, and untreated wounds of any kind. The GIs were issued brand-new clothing, and the old filthy uniforms—the same ones they'd been wearing since the fall of 1944—were burned. The men were also given vitamins, glasses of milk, good food, and cigarettes.

"A glass of grapefruit juice almost killed me," Paul Stern said. Death by overeating was no laughing matter. "Some kids got terribly sick because their stomachs had shrunk. In fact, one fellow at Lucky Strike gorged himself and he didn't make it."

The men liberated from Stalag IXA were among the first POWs to arrive at Lucky Strike, and, as Sonny recalled, they were met by "well-meaning Red Cross ladies who plied us with greasy donuts and coffee. Needless to say, we wolfed them down with little regard for what this diet was going to do to systems that had become accustomed to thin soup and ersatz bread. One unfortunate put away eighteen donuts and promptly died. In the next days, we were switched to eggnog and other more appropriate food."

Though Roddie, Lester, Skip, Paul, and Sonny gradually began to gain weight and regain their strength, other GIs were too far gone with malnutrition and disease. Hank Freedman didn't make the trip to Camp Lucky Strike with the other GIs but instead, upon landing at Le Havre, was taken to a "proper"— permanent—hospital. At five feet five, he weighed less than 110 pounds and would ultimately spend three weeks near death in the hospital in Rouen and then one week recovering in another hospital in Paris.

That group of nearly 1,300 noncoms were the first liberated American prisoners to arrive back in Allied control, and though they didn't think of themselves as heroes, they were treated with such enormous respect, Lester recalled, "our handlers seemed to feel that we were something close to it. . . . We were cleaned up, given new clothes, fed well and plenty, and treated like babies. We all got promotions and were given awards. I got three battle stars and ribbons and I was putting on weight very rapidly."

The stay at Camp Lucky Strike was brief—only four days—but few of the GIs would ever forget it. During their time there, on April 12, the radio abruptly announced that President Roosevelt had died in Warm Springs, Georgia, at the age of sixty-three. Few of the soldiers knew he'd been gravely ill for weeks; at Lucky Strike, most GIs stood in silence, some wiping away tears.

"Although I was born while Calvin Coolidge was president, and lived through the disaster of Herbert Hoover, my first awareness of matters political started with FDR," Sonny wrote. "In all those years of growing up, going to school and fighting a war, he was the only President I had known."

Lester felt devastated. "FDR was our hero," he said. "We didn't know what kind of president Truman would make. It was such a sad time."

Two days later, on April 14, 1945, first thing in the morning, the men piled again into trucks and drove the sixty miles back to the place they'd first come ashore: Le Havre Harbor. This time there would be no wading through breakers, getting soaked up to the armpits in frigid North Atlantic surf. The men boarded the USS *General W. P. Richardson*, a 622-foot troop-transport ship that had been in US Navy service only since October 1944. Frankie called it "the most beautiful ship" he'd ever seen. "It was spotlessly clean. Everything was pure white. It wasn't a liner, but it had the [RMS] *Aquitania* beat by plenty as far as comforts and accommodations go."

All the liberated POWs were sent below deck, where they discovered air-conditioned, roomy compartments. To soldiers now habituated to picking lice from their groins and underarms, and ignoring the stench of their buddies' urine and feces, the spotless cabins on the *Richardson* must have seemed a paradise.

Men who'd squatted over latrine pits festering with maggots now marveled at their floating accommodations. "The washroom was the best I'd seen since I left Camp Atterbury," Frankie wrote. "Clean porcelain wash basins, crystal-clear mirrors, showers, real commodes, and best of all, hot and cold running water." The group swapped stories of home, laughing as they enjoyed the simple pleasures of freedom again—shaving with sharp double-edged razor blades, using proper shaving brushes to lather up their faces, running water so hot it probably scalded their fingers.

Still anchored off the coast of Normandy, they lined up for a chow-time service fit for a king: boiled ham, baked potatoes, cabbage and carrot stew, and, according to Frankie, "all the good white bread you could eat, plenty of good fresh butter, and dill pickles and some damn good coffee." That fluffy white bread, so ordinary a commodity back in Brooklyn, the Bronx, and Knoxville, smeared in butter, was like some exotic delicacy served at the Ritz. GIs grabbed four or five slices at a time, clearly trying to forever erase the memory of the pitiful slivers of black sawdust bread on which they'd subsisted for months, and slurped down steaming hot cups of coffee, sweetened and lightened with cream, dispatching thoughts of that tasteless brown liquid the Nazis had passed off as coffee. "Boy, under conditions like this," Frankie said, "it won't be long before I forget all I've been through and it won't be long before I'll be good and fat."

The ship pulled out of Le Havre at 10:30 p.m. and headed first to England, to pick up some wounded GIs, before it would join a convoy of other naval ships making the still-treacherous passage back to North America.

Even in the short span of the English Channel, they ran into trouble—the previous night German submarines had been spotted. "Those dirty Jerries don't know when to quit," Frankie said.

The GIs watched as the crew prepared "ash cans"—Navy slang for depth charges—which they dropped overboard, detonated, and apparently got some U-boats.

"Trouble seems to follow me every place I go, and I still have a whole ocean to cross yet," Frankie jotted in his diary while sitting in his cabin. "But I'm sure we'll make it alright. God is watching, and you can bet I'm praying hard, how can I possibly go wrong if [I] trust in Him? Gosh, I won't feel really safe or happy again until I'm deep in Lucy's arms once more."

It wasn't simply the threat of roaming U-boats. In the English Channel, the captain of the *Richardson* slowed her to a standstill. The Nazis had done a thorough job mining those straits, and the navy crew began blasting away with their fore and aft guns at the mines floating in the water.

Groups of infantrymen howled mockingly at the sailors missing time and again with their powerful guns.

"Couldn't hit a mine for *sheet!*" one soldier laughed.

Another GI yelled up to the bridge: "Let me have a crack at
them!"

The infantryman shouldered an M1 Garand and in short order
cleared the water of any floating explosives. "*Cough, cough* went
the M1 and *blamb, blamb* went the mines," John Morse later said.

On Sunday morning, April 15, 1945, the men awakened to a familiar bugle call—reminding them of life on a "proper" military
base, like Fort Jackson or Camp Atterbury, rather than a miserable prison like Stalag IXA or a makeshift tent city like Lucky
Strike. They all washed with hot water, shaved, brushed their
teeth, put on clean uniforms, and sat down to a huge breakfast.
On the ship's deck, Frankie lit up a pipe and took in the sight of
the springtime coast of England. He said the rosary that morning, alone, in the absence of a priest.

After lunch, a large memorial was held in honor of FDR. Reports had come in from the States that a special train had returned
the president to Washington from Georgia, his casket carried on a
caisson in a military procession from Union Station to the White
House with a crowd of 500,000 people watching silently in the
April sun. After lying in rest in the East Room for a few hours,
and then a simple funeral service, FDR's coffin was taken back to
Union Station and placed aboard a train to be taken for burial at
his home in Hyde Park, New York. Lester, Roddie, and Frankie
stood in respectful silence, saluting when the flag of the *Richardson* was lowered to half-mast. The flag would remain lowered for
the rest of their journey home.

On Monday the sixteenth, the *Richardson* pulled into the port of
Southampton and took on the sick and wounded. There were men
who'd lost legs, men who'd lost arms, lost eyes. There were men

still severely shell-shocked, muttering to themselves, flinching, quivering, talking excitedly to shadows. "It made me feel lucky," Frankie wrote. Finally, with a twelve-ship convoy formed up, at six o'clock that same evening, the *Richardson* started for home.

Most of the men were too relieved—giddy almost—with the taste of freedom to worry about how dangerous the Atlantic passage might be. Even in those final weeks of the war, even as Soviet, American, and British forces raced toward Berlin, the Nazi *Unterseeboots* were still extremely active. The GIs learned via radio of a big symbolic victory that day: on April 16, 1945, US troops reached the Third Reich's spiritual heart, Nuremberg, the stage for massive Nazi Party rallies, including some of Hitler's most maniacal speeches and the announcement of draconian anti-Semitic laws. Hitler, from his deep bombproof bunker under the Reich Chancellery in Berlin, had ordered the city protected at all costs. When some German soldiers tried to surrender to Americans, waving white flags, they were mowed down by machine-gun fire from their fellow Nazis. The Nazis had no more manpower for the final defense of Berlin, and kids from Hitler Youth, as young as fifteen, were given rifles and grenades and thrown into the final, hopeless defense of Nuremberg.

Frankie confided to Roddie his fears that they'd come all this way, survived the Battle of the Bulge, the horror of the Christmas-time rail yard bombing, the hell of the camps, only to be sunk by a torpedo from some U-boat captain too fanatical, headstrong, or perhaps clueless to realize that the Nazi cause was lost.

Lester and others tried to push away any worries with onboard camaraderie, playing cards and rolling dice. There were well-organized games of craps on the deck almost all day long. Lester had his $100 mustering-out pay plus some back pay, and he got a hot hand with the dice, managing to parlay that few hundred into $2,000.

They weren't anywhere near US shores, but in many ways life on board the *Richardson* was starting to feel like home. In one compartment, the GIs had a crank-up record player, but only one vinyl record could be found on the whole ship: the Andrew Sisters' "Rum and Coca-Cola," the runaway hit of 1945. The strains of the sisters' three-part harmony and the infectious calypso remake echoed in the ship over and over. Soldiers and sailors alike jauntily walked up and down the decks singing:

Out on Manzanella Beach
GI romance with native peach
All night long, make tropic love
Next day, sit in hot sun and cool off
Drinkin' rum and Coca-Cola.

They weren't in the Caribbean sun, there were no beautiful "peaches" to chat up, certainly no rum-and-cokes to sip on, but at least the sun was out. Many men were sprawled out on the deck, tanning their faces and chests. The ship cruised along at a slow speed, and the sea was calm for hours. The men had no clue exactly where they were, but on April 17 they were told to set all their watches back one hour. "Ten more days of sailing," Frankie wrote in his journal, "five more hours to set back, and then America. Oh boy! Incidentally, I'm getting nice and fat!"

Roddie, Lester, and Frankie were awakened on the morning of April 18 to find they were in rough seas; up on deck, the air was cold and misty. They ate another huge breakfast, then smoked pipes up on the deck before going to a screening of *Star Spangled Rhythm*—the 1942 hit musical comedy starring Bing Crosby, Bob Hope, and Fred MacMurray—which most of the guys had seen before they'd even entered the army. Then a big lunch, another popular Hollywood movie, and Frankie noted, "I didn't eat supper. I'm not sick, but I felt that supper was all I needed to make me sick."

They could scarcely have imagined just a few weeks ago—staring at those miserable squares of black sawdust bread—that they'd soon be so full they'd refuse a plate loaded up with hot, rich American food.

For the next few days, the men remained anxious, but they kept playing craps, dancing, singing "Rum and Coca-Cola," and going to the onboard movie showings. Roddie and Frankie watched *Brother Orchid*, a 1940 American crime-comedy starring Edward G. Robinson, Ann Sothern, and Humphrey Bogart. "Old but refreshing," Frankie called the film, a lighthearted romp from the days when the US wasn't yet at war.

Each man also received his ration of ten Hershey's chocolate bars—an unthinkable luxury only three weeks earlier. From choppy seas, strong winds, and cold mist, Frankie and Roddie noticed that very soon the climate changed drastically, and by April 20 the temperature was so hot, the sun so strong, that the men wished for sunglasses and bathing trunks. They stripped off their uniform shirts and enjoyed the tropical sun, then went to watch a screening of *The Gang's All Here*, howling at the broad humor and letting out whoops of appreciation at Carmen Miranda.

They also heard the good news that on April 20 Nuremberg had fallen at last.

"A great birthday present for Adolf," said one of the GIs, smiling. Back in Berlin, it was Hitler's fifty-sixth birthday. The celebration, held in the Führerbunker, a claustrophobic bunker complex buried deep beneath the Reich Chancellery, was anything but festive. Most of the senior leaders of the Third Reich were there—Goering and Himmler, Goebbels and Von Ribbentrop, Keitel and Jodl—though many were secretly strategizing ways to get out of

Berlin. Some—as in the case of Himmler—were already in contact with the Allied commanders, trying to negotiate, trying to arrange a way to save their own skins.

The next morning, April 21, word spread quickly among the troops about why the climate was suddenly so tropical. The captain, in order to avoid U-boat activity, had plotted a course far to the south, and they were some 450 miles off the coast of Spain.

It meant more days at sea—more days postponing a reunion with loved ones. "But I don't mind going home the hard way, as long as it's the *safe* way," Frankie told Roddie and Lester. In his diary he wrote: "Another day gone by, and another pound of fat gained. I actually look human now."

On Sunday April 22, Roddie prayed from his Bible, and Frankie went to a rosary service in the morning with the other Roman Catholic GIs. Roddie and Frankie then found a trove of detective stories and enjoyed themselves, taking their minds off any worries about U-boat sightings or impatience about getting back home to solid ground. The sea turned unusually choppy, and the ship tossed about, Frankie said, "like a matchstick in a bathtub. It's a darn good thing I'm not allergic to seasickness."

The next day, April 23, was Frankie's birthday, but he was in no mood to celebrate. "This makes three years that I haven't had a birthday at home. That sure makes me mad. Just another day gone by. Gosh, but these days seem long and endless. Seems like we'll never reach home."

To ease the tension, Frankie and Lester sat through a double feature: *The Palm Beach Story* and *The Story of Louis Pasteur*. But no matter how they tried to distract themselves, the closer they got to home, the more many of the men began to feel a strange and unexpected sense of panic.

GIs who'd clung to memories of home during the Battle of the Bulge and months in the stalags now worried that the loved ones and the world they'd left behind had moved on without them, might not have a place for them. Roddie worried about trying to reconnect with his now ex-wife Marie. What if she wouldn't agree to reconcile? On April 25, Frankie wrote in his diary: "All day today I took to my bunk and thunk. Man, have I been thinking lately. I'm half-afraid to go home and I'm sick with worry. I keep thinking what if things have changed? What if there's trouble at home? What if something has happened? Will Lucy be the same or has she changed? That worries me more than anything. God. I wish I were home now."

The next day Frankie described his insomnia, fits of tossing and turning in the bunk next to Roddie: "I didn't sleep a wink all night long. Just can't seem to get Lucy and home off my mind. Well, anyway [I hear] we dock Saturday and leave the ship at eight o'clock. Saturday, please hurry up. It's just anxiety that's got me— nothing else."

Finally, after two weeks in the Atlantic, the coastline of New York came into view, white and red lights flickering through the predawn darkness and sea mist.

"When we reached the coastline that last night at sea, we saw the lights of Coney Island," John Morse recalled. "A tremendous cheer rang out. Home, at last!" In the morning, swarms of small vessels loaded up with civilians came out to greet the return-ing GIs. "We were met by tons of boats with people waving and cheering our return."

The USS *General W. P. Richardson* sailed into New York Harbor on April 28, and the men crowded the deck, straining to get a glimpse of Brooklyn and Manhattan—and when the Statue of Lib-erty came into view, her distinctive green-bronze figure bathed in sunlight, there was a deafening cheer. "We figured that we had helped keep her safe too," said Pete Frampton.

From the docks of Brooklyn, GIs were taken down to Camp

Kilmer at Brunswick, New Jersey, for another round of processing and debriefing.

The men liberated from Stalag IXA were one of the first large groups of returning US troops. "We were almost treated like celebrities," Lester later said. He was one of the ex-POWs who were asked to participate in a radio interview. "I called my parents: 'Tune in to this New Jersey station—I'm going to be interviewed.' After I was captured, they'd informed my family that I was missing in action and they didn't know if I was alive or dead until the month before I was liberated. The news didn't travel very quickly, so they were just overjoyed to hear my voice."

All the men were issued passes for sixty-two-day furloughs—a virtually unheard-of amount of time but an appreciation of the hard months they'd spent in Nazi captivity. All the sergeants and corporals were to have their teeth checked and cleaned; Lester and Paul Stern went together to a dental clinic on the Grand Concourse—that wide and impressive thoroughfare. Paul laughed as he remembered the Grand Concourse's nickname as the "Champs-Élysées of the Bronx"—it all seemed so different now that he'd paraded down the *actual* Champs-Élysées during the liberation of Paris.

After they'd both gotten a clean bill of health from the dentist, Lester smiled. "Paul, our apartment is just a few blocks further down on the Concourse. You want to come over and meet my mother?"

"Sure, why not? Let's go."

In their crisply pressed uniforms, nodding at passersby, they walked down the Grand Concourse to 170th Street. When they got out of the elevator and entered the apartment, they saw that Lester's mother, Frieda, had prepared a huge spread of food—corned beef and pastrami, fresh rye bread, kosher dill pickles, a dish of kasha varnishkes, fresh melon, and berries and honey cake. But the thing that leaped out at Paul was the fantastic platter of éclairs and bottles of milk.

They'd been in the US only a few weeks and Lester and Paul were still trying to gain back some of the 60 to 80 pounds they'd lost during the hundred-day ordeal at Stalags IXB and IXA.

"We sat there gorging ourselves on these delicious éclairs," Paul recalled. "After about ten minutes, the French doors to the dining [room] opened and Les's sister, Corinne, appeared. When I saw her, I fell in love. And I said to myself, 'That's the girl I'm going to marry.' At that very moment. And that's what happened. That was the moment I fell in love with her. And we were married two years later."

THOUGH HITLER COMMITTED suicide in his bunker on April 30, and the Nazi Reich was finished, the war still raged in the Pacific. Roddie, Lester, Frankie, Paul, Skip, Sonny, and Hank all realized they might be back in combat very soon, only this time fighting against the Japanese.

"I went to Lake Placid for two weeks of R&R," Lester recalled. "That's where Sonny Fox and I really bonded. We were two weeks at Lake Placid and then I was assigned back to Fort Benning, Georgia. He'd been promoted to staff sergeant. I was now part of a division that was being trained for the invasion of Japan. . . . It was pretty hot. We were battling on Iwo Jima, island-hopping, but everyone knew that in order to defeat Japan, we would have to invade the main island. MacArthur was getting together as many troops as possible to eventually make that invasion. I was part of that division."

Of course, neither Lester nor any of the other GIs knew about the Manhattan Project. But in August of 1945 Lester was given his final furlough before going overseas to the Pacific Theater.

"The time was too short to go home to the Bronx, so I went down to New Orleans," he said. "I was in New Orleans on August 5, 1945, when we heard the news about Hiroshima. And then the Japanese surrendered. Everyone was celebrating in New Orleans—it was like Mardi Gras in August. I was saved by President Truman deciding to drop those atomic bombs—for which he was widely criticized. I *never* criticized him. He probably saved a million lives—including mine. First Roddie. Then President Truman. I knew I was going to survive, come back, start my life."

On August 8, 1945, Lester was just twenty-two years old. "I'd finished college. I'd served my country. We were victorious. The Japanese were defeated. The Nazis were defeated. And life was beautiful."

For his part, my dad never spoke aloud about those early days of his repatriation to the States. But while in Stalag IXA, he wrote his final—prescient and expectant—words about a life of love, of friendship, of faith, he planned to lead from that day forward. "I am going back home to my relatives and friends," he wrote. "I feel as if I am going to be strange among them. I want no sympathy, I want peace, quiet, and more than anything I want God. I hope my actions won't cause the wrong feeling towards me. I don't want to do anything wrong, not the least little thing."

In the pages of the diary, he came to terms with how strange and incomprehensible his wartime experiences would be to his friends and loved ones. Even my dad, a man gifted with word-play, so gregarious and witty, found it impossible to express his precise feelings about everything he'd been through in the past six months.

"There is a lot more I would like to say, but I don't know how," he wrote in the final entry of his diary. "I have kind of poured my heart out here, and it is foolish I guess, but as I said at the beginning, I wanted to get it off my chest. I don't think I would ever have nerve enough to tell anyone this, but I feel as if I can let it be read. I would like to repeat, I have only been overseas a little less than six months, but I have got a good idea what combat is. You've got to be there to know."

PART VI

*"For I know the plans I have for you," declares
the* Lord, *"plans to prosper you and not to harm you,
plans to give you hope and a future."*

—JEREMIAH 29:11

TWENTY-FIVE

I N THE END, it all came back to the beginning—back to my
need to understand what my father had been through.

Though I could never comprehend the terror of combat or, as
my Dad wrote, know what it was like to "be there," I've certainly
tried to retrace his footsteps. By traveling across the country,
meeting the men whose lives he saved, by learning their own intimate experiences of the war, hearing their firsthand vision of
Dad, I started to recognize my father in their stories and see him
with new eyes.

Later, when I visited Europe and actually walked through the
same primeval-looking forests of Belgium where Dad had fought,
knelt down to feel the same damp earth that had erupted around
him so furiously in the predawn hours of December 16, 1944,
when I saw the positions he and the rest of the 106th Division had
tried so mightily to hold in the Ardennes, I felt as if I were rolling
back the decades. And as I started to solve the mystery that had
bothered me since Lauren came home with her college assign-

ment, I felt Dad once again standing over my shoulder, helping me fill in the missing pieces.

It was as if he were with me again.

One bright morning in Germany, walking up that steep hill from the Bad Orb train station to the site of Stalag IXB, I was stunned into silence at the sheer beauty of the surrounding wooded Spessart mountain range—and I felt instantly Dad's heartbreak and loneliness, felt it as clearly as I've ever felt anything in my life. The forested hilly German panorama he'd gazed up at so eerily resembled the Great Smoky Mountains of our beloved Tennessee.

The father I had known growing up had been so ordinary. To be sure, he had been full of life, and a sincere person of faith. But like most dads I knew, he had been common—even flawed. Yet my transformative journey had revealed something greater. While Dad's life had been marked by tragedy and hardships, it had been well lived and deeply felt. I know that now. My father wasn't perfect; none of ours are. But what I've learned is that you don't have to be perfect to do something extraordinary. Ordinary people are heroic, and an ordinary life lived well is, indeed, extraordinary. My father's story is a testament to that. And so are the stories of the men who served alongside him, the men my father saved in Stalag IXA. Like ripples in a pond, my father's actions all those years ago continue to resonate today, in unexpected ways.

Learning about Dad not only changed my conception of the world and the power of God's grace; it also transformed me and my family.

My father had seemingly lived several lifetimes before I was born. I knew he'd fought in both World War II and the Korean War before meeting my mom, but the details of his earlier years had been largely unknown to me and my older brother, Mike. In early January 2016, while doing research about my father's wartime experiences, I asked my mom what she knew about Dad's past. How much did she know about my father before his enlistment in the military? She casually mentioned that Dad had been married before he met her. Mom told me that my dad had married his high school sweetheart, Marie, during his army days but had gotten one of those infamous Dear John letters and then subsequent divorce papers just before shipping out to Europe. After his World War II service, he came back home to Knoxville, and though he made a concerted effort, he couldn't patch things up with his first wife. (Thankfully, he didn't give up on the idea of marriage; my mother and father were wed in 1953—and were a happy couple until his death in 1985.)

After my mom told me about Marie, she added—as if it were common knowledge—that I had a sister from that first marriage. A sister I'd never met. A sister who knew about me. A sister I soon discovered had lived in Knoxville all my life. A sister I had probably crossed paths with on several occasions when buying hot dogs at Paul's Market, a store she and her husband had owned during my college days.

On January 15, 2016, Regina and I met my sister, Priscilla Edmonds Davenport, and her husband, Paul, for the first time at the

Chop House in South Knoxville, not far from where Dad grew up and near where Priscilla and Paul still live. I knew immediately she was my sister. She was a petite, attractive, well-dressed, energetic lady—a devout Christian who had the unmistakable "look" of an Edmonds. Twelve years my elder, we bonded instantly. We're still close today.

Priscilla was born in Knoxville on January 30, 1945, just three days after Dad stood up to the German major on behalf of his Jewish men. I don't know if Dad knew that Marie was pregnant while they were getting a divorce. His wartime diary is so specific about the pain of the divorce, he surely would have mentioned if Marie was expecting his child.

Why didn't he pursue a more traditional father-daughter relationship with Priscilla? Dad visited her after World War II and even sent her a doll he'd picked up while serving in Korea—a possession she still cherishes—but Priscilla told me that her maternal grandfather didn't like Roddie and essentially forbade him from visiting her. Marie's father was a tough character in Priscilla's recollection. For whatever reason, he felt Roddie wasn't "good enough" for his daughter. Nonetheless, Priscilla grew up with a loving stepfather in the same South Knoxville neighborhood where Roddie and Marie had met.

By connecting with Priscilla, I began the process of healing what had been one of the most painful moments in Roddie's life. His divorce from Marie devastated him; he felt cut off from what he thought was going to be his future. "I had lost what I had wanted all my life, a home, a wife, and happiness," as he wrote in the POW camp. Because of my Christian faith—the faith I learned from my father—I'm quite certain he would be pleased to know that his daughter Priscilla and I have connected, have begun to heal that wound in a way that he couldn't when he was alive. Ultimately, Dad's brave actions in Stalag IXA not only saved his men but have led to repairing a breach in the family that he so desperately wanted to fix more than seventy years ago.

I immediately noticed something at the Chop House during that first meeting with Priscilla: as we talked about Dad and his past, Priscilla kept referring to him either as "Roddie" or "your dad."

"Priscilla," I said, finally. "He's our dad. I know you haven't experienced him that same way, as a father, but he is your dad too."

Of course, Priscilla had always harbored unanswered questions about our dad. I read her the passages in his diary describing his profound love for Marie. All her life, she told me, she'd wondered if her father had loved her. Hearing and seeing the words of anguish about his divorce, written while he was a POW, has helped her find some measure of comfort and healing.

Yet another stunning discovery, another life-altering fact that I never knew about my father before Lauren came home from college during her sophomore year with that history project, kick-starting this whole stranger-than-fiction journey. I'm not sure why Dad never told me or my brother about Marie or Priscilla. Perhaps, like his POW experiences, they were locked away in his vault of wartime memories, never to be seen or heard of again. Maybe those memories were simply too painful. Most likely it was because my father—like all the GIs I've met—enjoyed the days he was living and looked forward in hope to tomorrow. For me, meeting and getting to know Priscilla has been a gift, and I look forward to spending as many tomorrows together as siblings.

ANOTHER GREAT BLESSING has been meeting some of the men Dad saved. I've come to love these men and their families. Along with Dad, they are my heroes and great examples for us. They served our country well during World War II and overcame the same hellish POW camps as Dad. Their experience of the horrors of war didn't defeat them but rather inspired them to return to the United States, finish college, marry their sweethearts, raise families, and lead ordinary lives of extraordinary influence.

My dad's legacy lies in these men and their children, grandchildren, great-grandchildren, and the generations to come. One day, while I was sitting in the solitude of the Library of Congress, I began to calculate the full impact of my father's bravery.

By Lester's and Skip's reckoning, approximately two hundred Jewish infantrymen were saved that cold January Sabbath morning in Stalag IXA. And nearly thirteen hundred soldiers were saved two months later at liberation. Most of these men had children, grandchildren, great-grandchildren. I figured out that due to Dad's courageous leadership in the POW camp, more than *twelve thousand* people are alive today.

I made it a personal mission to sit with as many of the men Dad saved as I could. I wanted to learn more about what Dad had done in Stalag IXA, obviously, but I also wanted to hear and honor their own stories, to learn about the legacies they passed on. I am still on that journey.

In 2014, I drove down to meet with Hank Freedman in Suwanee, Georgia, about thirty-five miles north of Atlanta. Hank was ninety years old, living in a one-bedroom apartment near his loving family. He sat ramrod straight in his armchair, with a full head of white hair, strong and suntanned, wearing an aqua-blue short-sleeved shirt and neatly pressed khaki pants.

Hank told me about how the grandmother who had raised him fainted when he telephoned her from New York City on May 8, 1945—V-E Day—to say that he'd spent ninety-eight days in German POW camps. After the war, Hank enjoyed a successful business career, making frequent trips to the Far East as a buyer for Rich's department store in Atlanta. As a senior citizen, the once Yiddish-speaking Orthodox Jew accepted Jesus Christ as his personal Lord and Savior at Shadowbrook Baptist Church in Suwanee. He survived cancer and was married for fifty-one years to his lovely wife, Betty, until her death in 2004.

We talked all afternoon about his hellish experience and about Christmastime 1944, of being trapped in the locked boxcars on the siding in Limburg, Germany. Hank described again the nearly miraculous calm in the boxcar, how the terrified POWs stopped struggling and flailing when Roddie led them to pray.

Off to the side of Hank's living room was a small study, which Hank called his "prayer closet." There, he told me, "I study the scriptures and pray daily." The walls were filled with his World War II medals and badges, with framed awards—one from the French Foreign Legion—and memorabilia of his family and career.

He told me how much Roddie's demonstration of faith in the boxcar had influenced the rest of his life—a life, I can say, that had been well lived.

"It had an effect on me," Hank said. "Here it is more than seventy years later, and I remember that incident very vividly. Roddie's faith amazed me. It was my first seed of faith."

Hank's decision to become a born-again Christian was just one of the extraordinary acts of faith I discovered while on this journey.

ANOTHER REMARKABLE ACT of faith was one I heard about from Paul and Corinne Stern, and from their daughter, Joanne, during my visits to their home in Reston, Virginia. Among all the World War II memorabilia Paul had in his den, there were several pieces of fragile yellowed paper—the official notifications that Paul's mother, Jennie, had received, first that Paul was "missing in action" and others confirming that he was a prisoner of war.

Upon reading the telegrams, Jennie made a solemn vow: she would observe the Sabbath, strictly, from that moment on. That meant she would do no work, wouldn't turn on a light—from sundown on Friday when the Shabbat commenced until sunset on Saturday she would observe God's day of rest.

And should her son survive the ordeal, she promised, she

would "remember the Sabbath day, to keep it holy," for the rest of her life.

When I was visiting with Paul and Corinne in Virginia, their daughter, Joanne, told me how deeply her grandmother's decision had affected her as a twelve-year-old girl. "There isn't a week that goes by that I don't think about my grandmother's commitment—when she received the telegram saying that her son was missing in action and subsequently found out he was a prisoner of war. She made a commitment at that point to observe the Sabbath, which included not driving in a car. Or even carrying an umbrella. I remember the night of my bat mitzvah—a Friday evening, we were all going to the synagogue and it was raining cats and dogs."

Corinne was driving to the Temple Sholom a couple of miles from their home in Westbury, New York, with young Joanne in the passenger seat, but Jennie insisted on walking, and her son, Paul, chose to walk with her. "They kept walking in that downpour, in the pitch-dark, all the way to the synagogue. Every few blocks my mom would stop the car and my grandmother would get in—we had towels and we would dry her off. I wondered: Is she really going to go back out again? And she did. She went right back out and continued to walk to the synagogue with my father by her side. I vividly remember the beautiful blue silk suit that she wore—and being so appreciative that she'd kept her commitment, had such discipline, and most importantly that she was able to attend my bat mitzvah."

"Did she have other grandchildren?" I asked.

"Many other grandchildren," Joanne told me, "and there were a few, unfortunately, whose bar mitzvahs and bat mitzvahs she could not attend because the synagogue was too far for her to walk and there was no hotel nearby the synagogue where she could have stayed. Others might have just said, 'Well, that was a long time ago, that was 1945, and things have changed,' but she didn't. She honored her commitment."

Paul sat there smiling—he remembered that rainy evening well. "We walked down Bromton Drive."

"Yes," Joanne said. "Bromton and Stratford Drive."

What about the story, I wondered, had so resonated with Joanne that she thought about it "every week" of her life?

"In this generation, honestly, Chris, I think we've lost the definition of commitment and discipline."

I agreed. "I deeply admire a mom, having her son lost in action overseas, crying out to God, making a vow and then living that commitment."

"My mother had three sons overseas," Paul added.

"And what's important is teaching it to children," Joanne said. "I've shared this story with my own children over and over again, as an example of the kind of commitment and discipline my grandmother had."

On March 30, 2017, at age ninety-three, Paul passed away in Virginia surrounded by his loving family. It was seventy-two years to the day after his liberation from Stalag IXA. Paul's last message to his children and grandchildren was about love: with pen and paper, he drew a heart to illustrate his desire for them to remain close, asking them to make his treasured retreat in East Hampton, Long Island, a place to love and share with one another.

LIKE PAUL, SKIP FRIEDMAN became another friend and part of my new extended family. Skip was committed to social justice—and it must have stemmed, in large part, from his POW experiences. Even when I met him, in his nineties, Skip was such a vibrant character—in my eyes, the modern definition of a Renaissance man. In addition to having been a star football player, avid historian, and accomplished lawyer, he could act and sing. (Like Paul and Lester, Skip loved the heyday of Broadway musicals: *Oklahoma!*, *The Music Man*, *Fiddler on the Roof*, and *South Pacific*.) After the war, Skip sang in a barbershop quartet and even appeared as a spear carrier in a production of *Aida*.

Skip's faith differed from his upbringing—as an adult, he had come to see the strict Orthodox Judaism of his parents and grandparents as rather extreme and chose a more moderate path; he and his wife were active in their local Reform synagogue in Shaker Heights, though Skip's children say their father wasn't especially devout.

Nonetheless, like my dad, Skip possessed profound moral clarity and was a man utterly without prejudice. When Skip's daughter Amy was a senior in high school, she became best friends with a student named Gabi, visiting from West Germany. Amy asked her parents if Gabi could move into the family house in Shaker Heights for a year. "After she was living with us," Amy said, "we all learned that Gabi's father had been a German soldier in the Second World War. And some of my parents' friends gave them grief," but Skip dismissed them with a muttered burst of GI profanity. "He didn't care what anyone thought."

During a recent conversation, Skip's son, Peter, told me about his father's love of the song "You've Got to Be Carefully Taught," from the classic 1949 Rodgers and Hammerstein musical *South Pacific*. Peter said, "My dad particularly loved to sing those words: 'You've got to be *taught* to hate and fear . . . you've got to be carefully *taught*.'"

My friendship with Skip was remarkably warm and intimate—I met him so late in his life, but he instantly accepted and loved me. I'll never forget his wit, intellect—even in casual conversations he spoke in philosophical terms, leading us to share some rich words about the value of life.

In their retirement years, Skip and Penny—who'd been a social studies teacher at the local high school across the street from their house—traveled the world, enjoyed tennis and music, and were active in their local synagogue and numerous political groups, including the League of Women Voters. Plus, Skip served on the boards of several hospitals in Ohio. He doted on Penny through her battle with dementia—rather than hospitalize his wife, he cared for her, an almost full-time job—until she died, after more than sixty-five years of marriage. As a widower, Skip continued with astonishing zeal: still practicing law and enjoying the company of his ten grandchildren. He cherished each day and touched countless others with his brilliant mind, optimism, and generosity. Cared for by medical staff of the Cleveland VA hospital, Skip died peacefully on February 23, 2017, in his Shaker Heights home, at age ninety-two, surrounded by his loved ones, stacks of his books, and photos of his beloved Penny.

IN LATE JUNE 2018, I met Frankie Cerenzia's daughter, Lorna Nocero—she joined Lester, Regina, and me for a lunch at the Harvard Club. I'd been searching online for records of the men in my dad's journal and found that a "Frank Cerenzia, Staff Sergeant, World War II" was buried in a veterans cemetery in East Farmingdale, Suffolk County, New York. Luckily, "Cerenzia" is not a common surname, and through some ancestry detective work, I found records of Frank and Lucille and the married names of their daughters. It was a cold call—to a high school on Long Island—but it turned out that I'd hit the bull's-eye: Lorna was indeed Frank's daughter.

Lorna had brought with her a folder of letters, pictures, news clippings, and—amazingly—the diary her dad had kept in the POW camps: the exact same type of notebook my father had used. We examined it gingerly, like a museum piece, at the Harvard Club.

"Does your father's book look like this?" Lorna asked.

"*Exactly* like this," I said, astonished, as I held Frankie's diary.

"Where did they get the books?" Regina asked.

"The International Red Cross was giving them out," I said, carefully cracking open the pages, each one filled with Frankie's seventy-three-year-old script, each day of his captivity meticulously dated. The words were so clear, it seemed as if they'd been written yesterday.

"Your father's got great handwriting," I said, then in the middle of one page, a passage leapt out at me. "Look at this: 'On the 18th of January the Germans showed their true colors. They segregated all the Jews from the gentiles.'" On another page of the diary, Frankie had described my own father—and how at Bad Orb a group of the GIs had refused the illegal German order to provide information about religion and instead stood outside, stripped naked to the waist in the freezing weather: "'Roddie Edmonds was also one of us,'" I read aloud. "'Roddie and I have been together so far since the very beginning of my army career up until now. We will continue to be very close to each other even after we've been discharged and sent home. He lives in Knoxville, Tennessee.'"

Of all the guys, Frankie's life after the war was the shortest. Lorna explained that her dad—a lifelong cigarette smoker—developed heart disease while still in his forties. In the days before angioplasty or bypass surgery, there was little Frankie's doctors could do. He suffered a massive heart attack and died, age forty-nine, in October 1973, when Lorna was just twenty-two. Lorna got wistful when I asked what she knew about her father's younger years. She said she knew painfully little. "I feel like he died before my life even began."

Sitting at a round table in the Harvard Club's dining room,

we mirrored each other: I was so curious to know more about Frankie, and she was curious to know about Roddie. There we sat, exchanging stories, mementos, favorite moments, as Lester smiled, sipping coffee.

Frankie had returned to Brooklyn straight after discharge from the army. He and Lucy were reunited and lived in the Marine Park section of the borough. He became a typical "hardworking, breadwinner" Italian American father, Lorna said, and made enough money selling insurance that he was able to build a nice home in Howard Beach, Queens, in 1966, where Lorna and her siblings were raised. Frankie had lived only seven years with Lucy and his daughters in their dream house before he passed away. Lorna still lives in it with her husband and children.

She was fascinated to learn that Lester had been part of the "threesome" from the Fort Jackson and Camp Atterbury training days and had known her father in his youth.

"Oh yeah, Frank was in my squad," Lester said. "We were assigned to the same jeep. I think there's a picture of us together."

"After the war, did you ever meet him in the city?"

"No," Lester said. "We got separated after the prison camp. I went to Fort Benning and he went somewhere else. But we were good friends—army buddies. He was a great guy, your father."

"Yeah, he was a great guy." Lorna smiled. "As an insurance salesman, all his clients loved him. They'd call him up for advice on all sorts of issues other than insurance. People just really loved him."

Like Roddie, Frankie had never talked to his family about his wartime experiences, and I wondered if Lorna had noticed anything—any sign—that he'd gone through such trauma in the Battle of the Bulge or in the two Nazi POW camps.

Lorna shook her head resolutely. "My father was the jolliest man," she said.

"Mine was too."

"He never seemed like he was in a bad mood," she said. "My mother would be grumpy, but my father was always chipper, always positive. If you came to him with a problem, he always had a positive attitude."

It was another common thread I'd noticed, with my dad and many of his men: never taking tomorrow for granted, always living in the present.

"Lorna," I said. "Your dad reminds me of Skip Friedman. He told me, 'Chris, since I left that POW camp, I've *never* had a bad day. Not one—ever.'"

"I'm sure they just appreciated being alive," Lorna said.

"Actually, Skip told me, 'We *died* in that camp and we were reborn.'"

As our dinners arrived, Lorna was curious to know how I'd managed to find her.

I smiled. "Well, I work for God, so I'll give him the credit."

A FAMOUS PASSAGE in the Talmud teaches us: "Anyone who saves a single life, it is as if he saved an entire world."

As a Christian, I believe that with all my heart.

That's why I left my full-time job to focus on spreading Dad's message, telling the world about the transformative power of selfless sacrifice and moral courage.

In the library of the Harvard Club, Lester had asked me, "Chris, do you know your congressman in Tennessee? You should tell the story to your congressman and see if he agrees about the Medal of Honor. If he does, then you should seek his help."

As I drove back home with Regina and Austin, I realized that my life had taken on a new direction, a new mission: I had the opportunity—indeed, the obligation—to inspire people everywhere with Dad's story.

In Tennessee, my congressman, Jimmy Duncan, heard the account of my father's heroism and agreed that Roddie Edmonds deserved the Medal of Honor. He asked Tennessee senators Lamar Alexander and Bob Corker to join the effort. An entire team of Tennesseans went to work gathering affidavits from eyewitnesses, death certificates of superior officers, historical information regarding the camp, actions taken, eyewitness testimony—all to help recognize what my dad did.

While I traveled around the country to meet the men Dad saved, Lester remained actively involved in the Medal of Honor efforts. He often said that the "defining moment of his life"—from the military through Harvard Law School and beyond—stemmed from Roddie's actions on that bitter-cold January morning in Stalag IXA.

"The lesson of that day has shaped my life," Lester wrote in an affidavit for the Medal of Honor recommendation. "There have been times when you must take a calculated risk, however perilous, to stand up and do the right thing for yourself and those for whom you have responsibility. Roddie's courage influenced me to attend Law School with the help of the GI Bill. One man's courage saved many lives, mine among them. When I look back at all the years since that fateful day, I find many occasions in my personal, family, and professional life when I can link my decisions and actions to my service in the war and to that experience when I watched Roddie standing up to the Nazi major. I am still doing that now—at age ninety."

Advocating for my father's recognition became a personal mission for Lester. He wrote to the New York congressional leaders, enlisting their support. He also got the help of one of his closest friends, Larry Goldstein, an energetic businessman who—as I often say—likes to get things done yesterday. Larry frequently asked me for the latest information I was gathering about Dad. He followed up with calls and emails, asking for more clarifications, more details.

What I didn't know was that Larry had secretly recommended my father to Yad Vashem for consideration as Righteous Among the Nations. If he were to be selected—a long and difficult process—Larry wanted it to surprise me.

Not only did Larry submit copious documentation, he and his wife, Barbara, even traveled to Israel to follow up with key decision-makers at Yad Vashem. Both have become dear friends.

I later learned how exhaustive Yad Vashem's vetting process is—they spent hundreds of hours researching the information, interviewing POWs, gathering evidence, and verifying the specifics of the Stalag IXA rescue story.

On June 6, 2015—the anniversary of D-Day—I received an unexpected phone call from the Consulate General of Israel in Atlanta, relaying the news that Yad Vashem, Israel's national Holocaust memorial, had decided to honor my father for extraordinary bravery in saving Jewish GIs during World War II.

The consulate also informed me that my dad—Master Sergeant Roddie Waring Edmonds of Knoxville, Tennessee—would be honored in Jerusalem as one of the Righteous Among the Nations, the highest honor a Gentile can receive from the State of Israel.

Only four other Americans have received this honor. No other US serviceman ever has.

IN MID-DECEMBER 2015, I made my first visit to Israel. As a pastor, visiting the Holy Land was the fulfillment of a lifelong dream. I'd come to Jerusalem not only to honor Dad but to study the Holocaust with Christian leaders. Standing at the gates of Yad Vashem, the World Holocaust Remembrance Center, I began reading Isaiah 56:5:

*And to them will I give in my house and within my walls a memorial and
a name that shall not be cut off.*

I could scarcely believe I was here. Transfixed by the passage
of scripture, my thoughts took me back seventy years, to that enraged Nazi major with a pistol pressed to Dad's head.

"Sergeant, you will order the Jews to step forward or I'll shoot
you right now."

You should not be here, my thoughts whispered. *Your father should
have died. He should have been shot and the Jewish Americans should have
been sent to their death. The Nazis should have never given in to your
father.*

But the Nazis *did* give in—not once but twice. In the waning
days of March 1945, when all the American POWs were ordered
to form up outside the barracks to await orders to march out, my
father, leading a revolt, ordered all American prisoners not to co-
operate with the German orders. Dad knew it was imperative to
their survival that they stay in the camp. The POWs had been in a
state of starvation for more than one hundred days, and Dad knew
many of the men could die from a forced march or in the crossfire
of battle.

Reading Isaiah 56 again, I brushed a tear from my cheek.

I'm alive. I'm more alive than I've ever been.

As I ran to catch up to my group, I thanked the Lord for saving
Dad and his men and for giving my father the moral courage to
be as "bold as a lion."

"A 'Righteous' is a non-Jew who helps Jews—someone who decided to leave the position of the bystander, someone willing to
pay a price," said Irena Steinfeldt, Yad Vashem's director of the Department for the Righteous Among the Nations. "Someone who
risked himself in order to help a Jew . . . someone who was willing
to share the fate of the Jews."

Irena explained: "It's an attempt to recognize that every person
is responsible for their deeds. Every person has a choice between
good and evil." There are some twenty-six thousand such Righteous Gentiles honored at Yad Vashem, but the case of my dad was
entirely without precedent: "It was the first time we received a request to honor an American for saving American Jews. Although
[Roddie] is a POW, although he is already in a very dangerous position, he decides to risk himself in order to save his fellow POWs.
He didn't have to do that. He could have said 'I have paid my dues;
I have done everything for my country, I'm a patriot, I served my
country well.' Nevertheless, he went the extra mile. It's a story
that shows you, in whatever situation, wherever you are, no matter how bad it can be, a person can always make a difference."

Being honored among the righteous is a fitting tribute to Dad,
a sincere Christian who expressed an infectious love for everyone.
He joins a group of ordinary people who mustered extraordinary
courage to uphold the goodness and dignity of humanity. In a defining moment, Dad stared evil in the face, refused to join the
masses, bowed to no one, and chose what was right—regardless
of the risk, regardless of the consequences.

Of course, my family and I are very proud of my father. We're
proud of the decisions he made and the life he lived.

RECENTLY, I'VE TRAVELED across the US sharing Dad's remarkable story. I visited Europe for the first time, traveling with a film
crew from the Jewish Foundation for the Righteous to retrace
Dad's wartime footsteps—from the Belgian battlefields to the sites
of the German POW camps. I visited leaders in the Knesset and met
privately with the prime minister of Israel, Benjamin Netanyahu,
in Israel. In late 2018, I was invited to speak about Dad's actions at
the European Parliament in Brussels. When my father received the
Medal of Valor from the Simon Wiesenthal Center in Los Angeles,

I shared the details of Dad's story with an audience of Hollywood stars, including Barbra Streisand, James Brolin, Michael Douglas, Norman Lear, Tobey Maguire, and Ron Howard.

But I've also been deeply touched by how my father's story has affected other people, men and women, and boys and girls, not directly connected to it, anonymous faces in the crowds I regularly address.

Last year I received a letter from a woman named Cathy of Portland, Oregon, who'd heard about me through her local synagogue's newsletter—their rabbi had urged the entire congregation to learn Dad's story as they prepared for the upcoming Passover holiday. Cathy wrote that online she'd heard me speak and watched a documentary of me retracing Dad's footsteps in Europe, which included interviews of Lester, Skip, Paul, and Sonny.

"I remember I was sitting in my kitchen weeping while one of the men struggled with his emotions as he bestowed upon your father the title, 'a righteous man.'" Cathy wrote that she'd been raised "in an Irish Catholic household in Chicago, in a liberal home where we marched for Civil Rights and watched the television news showing the treatment of blacks in the South and at tender ages read 'Letters from a Birmingham Jail' and *To Kill a Mockingbird*. Chicago at the time was the most segregated city in the Northern Hemisphere; we were no strangers to racism."

As a young adult Cathy had studied all the books she could to understand "the deep stain of slavery on our country," and after college she "fell in love with and married a Jewish man. I read dozens of novels and histories about Jews, and became deeply involved with Jewishness and gravitated toward Judaism. We raised our three sons in those traditions, and I am at home worshipping in the progressive Reconstruction branch of Judaism." She'd been deeply moved, she wrote, by the "true story of someone risking his life rather than turn over the Jewish soldiers to the Nazis."

But then something wholly unexpected happened to her.

"Despite my view of myself as a loving person opposed to racism and anti-Semitism and committed to 'tikkun olam [heal the world],' I myself have been prejudiced—against white Southerners. Especially pastors, whom I stereotyped as hypocritical and racist. I had never really faced this." Now nearly seventy, Cathy said that she'd harbored this animosity against men like me for a long time. "To see your dad, how young he was, what he did with no hesitation—it's rare to ever witness that level of love and bravery. Your dad's story made me realize that I had created my own 'other' to dehumanize and look down on. It was just like a weight was lifted off of me. . . . Your search for what happened to him, your discovery, and your work [to] share that story—I hope you know how much that matters. The fact that your father risked his life with his testimony that "we are all Jews" moved me, but I am also grateful that the story of one man—Roddie Edmonds—cracked me open to my own blind prejudice." Cathy said she'd been sharing Dad's story with her friends and family, and also sharing with them how it changed her personally.

This is, indeed, the most unlikely—and deeply appreciated—aspect of my father's legacy. I'm constantly amazed by how Dad's act of moral courage, three-quarters of a century ago, still resonates with and inspires people like Cathy.

All of this would no doubt confuse Dad, who probably thought he was just doing what he had to do. If he were alive today, he would ask, "Son, what's all the fuss? I opposed the enemy, I protected my men, I honored God. I was just doing my job."

In fact, that's what President Barack Obama said at the Embassy of Israel in Washington, DC, January 27, 2016. It was International Holocaust Remembrance Day and the seventy-first anniversary of the liberation of Auschwitz; President Obama was speaking at the Righteous Among the Nations award ceremony, introduced by filmmaker Steven Spielberg.

"I know your dad said he was just doing his job, Chris, but he went above and beyond the call of duty," the president said. Looking me in the eye, he added, "Faced with a choice of giving up his fellow soldiers or saving his own life, Roddie looked evil in the eye and dared a Nazi to shoot. His moral compass never wavered. He was true to his faith, and he saved some two hundred Jewish American soldiers as a consequence. It's an instructive lesson, by the way, for those of us Christians. I cannot imagine a greater expression of Christianity than to say, 'I too am a Jew.'"

President Obama raised an excellent point that day. Dad's actions were a true manifestation of Christianity, a true expression of Christ's love for humankind.

Dad loved all his men. As Lester Tanner said at the Righteous Ceremony, "Roddie could no more have turned over any of his men to the Nazis than he could stop breathing."

To Dad, his actions weren't an act of valor; they were simply the right thing to do.

Faith, and his military training, had given Dad a deep sense of right and wrong. His moral clarity gave him focus to see evil for what it was, and the courage to stand firm—even with the Wehrmacht major's Luger pressed to his forehead.

Under the harshest conditions, my father and the other POWs discovered that life is a precious gift. "Son," my dad would often tell me, "never forget: we are fearfully and wonderfully made." From

their day of liberation forward, nothing was taken for granted: each breath of fresh air, each morsel of bread, each cup of coffee was something they cherished.

Today, even in their ninth decades, guys like Lester—he's now ninety-five and vowing he'll see one hundred—awake each morning with a sense of profound appreciation and joy.

When Dad gave the order for everyone to fall out, any one of the men could have refused. Any number of the Americans could have simply stayed in the barracks where it was safe. But they didn't. They all fell out, risking their lives. Any one of them could have given in to fear and turned over the Jewish Americans in their ranks. But they didn't. They understood that life was about *all* of them.

I'm often asked why I think Dad did what he did, why he stood resolute even when faced with the threat of point-blank execution by Major Siegmann. Lester feels it was my father's desire—well, his *need*—as a soldier to resist the humiliation of captivity, to fight back against the Germans in any way he could. It was my father's

way of remaining a proud US infantryman even without his beloved M1 rifle.

I wish my dad were around for us to ask him. But I can almost
hear his answer in a recording made, sometime in the 1950s, of
Dad singing one of his favorite gospel songs by Albert E. Brumley:

"I'M A PRIVATE IN THE ARMY OF THE LORD"

Jesus is my captain and he
leads me all the while,
leads me all the while,
leads me all the while.
I am not a hero, but I'm in the
rank and file.
I'm a private in the army of
the Lord!

If he were by my side today, if I could ask him why he did what
he did in Stalag IXA in January 1945, I'm nearly certain what his
answer would be: my father was willing to die to save Jewish men
under his command because he believed a Jewish man—Jesus
Christ—had died to save him.

Growing up, of course, I had no clue that my dad had risked his
life to save 200 Jewish American GIs. I had no clue he'd risked his
life to save nearly 1,300 starving infantrymen of all faiths and eth-
nicities. As a kid in Knoxville, my dad was just the jovial guy with
the booming baritone voice, belting out spirituals in the church
pew next to me, the volunteer coach teaching my buddies and
me how to hook slide into second base—he was simply "Dad,"
chomping on one of his Dutch Masters "chewin' seegars" as he
mowed our lawn on steamy Tennessee summer evenings. But he
was no less of a hero in my eyes.

I guess that's what's most remarkable about my journey to discover what my father did in the Second World War—the realization that any *one* of us has the untapped potential to do something incredibly courageous.

Not a day passes for me now when I don't marvel at this epiphany: it could be the most unassuming person—a grandmother wheeling her cart through a supermarket or a teacher welcoming her students with a smile, a teenage boy hunched over a book at an airport gate, or a young soldier in harm's way—we all have the potential to change the world simply by standing up for what's right.

True heroes, I've learned, are rarely the larger-than-life characters of comic books or Hollywood blockbusters. They walk among us—like my dad did—virtually unnoticed, every day. They make the world a better place, quietly, anonymously—one person, one action, at a time.

Afterword

MY FASCINATING JOURNEY continues. I hope it never ends. It's been a journey of life and love guided by the good Lord above. Without the small miracles of God's providence, and hundreds of helpful folks along the way, Dad's remarkable story—and the stories of the heroic men who served with him—would have remained lost and forgotten. I remain grateful that these stories exist to inspire thousands of people around the world.

I'm blessed that Lester Tanner, a paragon of heroism, has become my friend and mentor. His practical wisdom and positive spirit make him a natural confidant. I'm certain my father saw the same qualities in him.

When I first met Lester, he was eighty-eight, full of energy, and still practicing law part-time. I found a man who, like my dad, loved life—he never took a single day for granted. His zest for living reminded me of Psalm 118:24:

This is the day that the LORD has made;
let us rejoice and be glad in it.

To celebrate his eighty-eighth birthday, according to our mutual friend Larry Goldstein, Lester picked up Larry in his brand-new red Audi convertible and tore out of Manhattan, pushing the car's speedometer toward his age. "Before I knew it, we were crossing the George Washington Bridge," Larry told me. "We sped there in twenty-five minutes and I was a nervous wreck. Now we're on the New Jersey Turnpike and, yes, Lester is doing eighty-eight, weaving our way around the slow drivers in the fast lane. My prayer life caught up in short order."

Lester has lived well. His indomitable spirit, positivity, and generosity have influenced me and many others. Like all the men I met who served with Dad, Lester has helped and inspired countless people in his legal career and everyday life. In 1992, Lester won a landmark case against American Airlines and five of its executives, representing former executive Barbara L. Sogg in her gender discrimination suit. Following seven years of litigation, Lester's firm, Tanner Propp Fersko & Sterner, won the case, and a $7 million award for Ms. Sogg. "The jury has sent a very clear signal to corporate America that the public will not tolerate discrimination in any form," Lester told reporters after the verdict.

Lester's remarkable career—indeed, his entire life—has been the result of both hard work and ethical choices. Lester told me that what my father did back in Stalag IXA was a "life-changing experience," one that made everyone feel brave.

"On that day in 1945, I made a promise to myself to always do the right thing," he said. "Particularly when you're an attorney and you must make moral decisions, you have to remind yourself to take the risk, to do the right thing." Then, with a wry smile, he added, "For a New York City lawyer, that's not always the easiest thing to do."

At age ninety-five, Lester continues to keep that promise.

The journey that began around a kitchen table with my daughter led me to other kitchen tables, into other living rooms of men who still held on to vivid recollections of what they went through all those years ago. And just like my dad, few had ever shared those experiences with their families. We shed lots of tears, shared many poignant moments in those homes. Over time, they became family.

One indelible memory was seeing the daughter of one of Dad's fellow POWs crying softly as her father related his story. His eyes were distant; it was almost as if he were back at the camp. She said that she'd heard a few things from her father about the war, but she had never heard the heartbreaking tale he was telling me.

Sonny Fox, pioneering American television host, executive, and broadcasting consultant, has become yet another invaluable new friend. Regina and I first met Sonny—most famous for hosting the children's program *Wonderama* from 1959 to 1967—in New York. Once again, we were back in the Harvard Club, a guest of Lester. It was September 2015. Lester was glad to arrange the luncheon as Sonny, age ninety, was in town on business. Sonny, still tall and lean, sported a white suit with a crisp, burgundy-striped oxford shirt, which accented his full head of silver hair. He brought me a copy of his memoir, *But You Made the Front Page! Wonderama, Wars, and a Whole Bunch of Life*. Most of what I wanted to know about his time in World War II, he told me, was covered in his book. Sonny was the consummate entertainer, quick-witted, full of clever ad-libs and good-humored jabs. With Lester as his straight man, Sonny reminisced about their time during the war, weaving in one-liners and anecdotes that had both Regina and me howling with laughter.

Later, we had the pleasure of sitting with Sonny and his beloved wife, Cely, in Los Angeles when Dad received a Medal of Valor from the Simon Wiesenthal Center along with the late Israeli prime minister Shimon Peres and Christian humanitarian Johnnie Moore. And earlier this year, I visited Sonny at his home near Beverly Hills. As always, he was a convivial host, a natural-born raconteur. I'll always cherish my friendship with Sonny and Cely, who passed away a few months before that visit.

One deep regret I've felt is not having the opportunity to meet more of these heroic veterans—men like Jack Sherman and Ernest

Kinoy. I was blessed to talk to Jack on the phone in September 2016, but before I could visit him in his hometown of Rochester, New York, he passed away in early 2017.

I'll never forget his first words to me.

"It was so good to see your father's picture in the paper," Jack said, referring to recent news accounts about Dad's receiving the Righteous Gentile honor at Yad Vashem. "I said, 'Wow, I *know* that guy.' He and I were good friends. But I have one distinction." I waited expectantly. "No man ever *slept* with your father but me." We both had a good laugh at that.

Then, more somberly, Jack recalled how my dad and he had to huddle together in those pyramids of bodies, in the Bleialf churchyard, those first dreadful nights of capture, holding each other close just to keep from freezing to death, wondering if their Nazi guards might suddenly open fire with their machine guns. I still don't know why Jack and Dad were separated at the rail yard in Gerolstein, but Jack's POW journey took him eastward into Prussia, to Stalag IVB in Mühlberg, where he survived six grueling months before finally being liberated by the Red Army on April 23, 1945.

As we finished our conversation, Jack recalled my father's character.

"Roddie was very tolerant of everybody," he said. "That's the kind of person he was. And you can rest assured that his religion did play a part in his personality. Chris, your father was loved by everybody. Sure, he was a tough sergeant, but he had a heart of gold."

Jack did too. After graduating from the Wharton Business School, Jack was the owner of Sherman Battery & Auto Parts until 1997. In retirement, he delivered Meals on Wheels and regularly volunteered at the Jewish Home of Rochester. His sixty-five-year marriage to his beloved wife, Marcia Lou (or "Lulu"), was perhaps his proudest achievement. He was a devoted husband, father, and grandfather, who enjoyed classical music and the Boston Red Sox. Talking with Jack, and hearing his stories about my father, has been one of the highlights of my journey.

Ernest Kinoy is another remarkable World War II veteran and ex-POW I wish I could have met. Ernie—as all his friends called him—went on to have an accomplished career as a screenwriter, playwright, and president of the Writers Guild of America (East). He was a staff writer at NBC during the "Golden Age" of television in the 1950s, then won two Emmy awards, one for writing the 1964 "Blacklist" episode of the CBS drama *The Defenders,* and another for writing the second episode of the ABC miniseries *Roots* in 1977. Ernie had been locked in the same boxcar with Dad and Hank Freedman that terror-filled Christmas Eve of 1944. Somewhere along the journey, Ernie signed my dad's journal and, later, after being separated with the other Jewish privates at Bad Orb, miraculously survived both starvation and German brutality in the Berga slave-labor camp.

Connecting with my sister, Priscilla, has been an unexpected miracle. Just two weeks after our first meeting in Knoxville, Priscilla and her husband, Paul, joined us for the Righteous Ceremony in Washington. It wasn't easy for me to get them invitations, but it was important to Priscilla—and to me—that she be there. It helped her to see our father more clearly and contributed to her emotional healing. She and Paul, like everyone attending, were moved by Dad's selfless acts of bravery. For me, it was a blessing to have my newly discovered sister there. The desire Dad had expressed so long ago to heal his family and to have a relationship with his daughter was in the process of being fulfilled.

Recognized along with Dad posthumously at the Righteous ceremony in Washington were Lois Gunden of Goshen, Indiana, and Polish citizens Walery and Maryla Zbijewski of Warsaw. Gunden, a French teacher serving as a Mennonite missionary, established a children's home in southern France where she sheltered Jews she'd helped smuggle out of a nearby internment camp. She even protected the children when French police showed up at the home. She was later detained by the Germans and released in a prisoner exchange. The Zbijewskis hid a Jewish child in their Warsaw home until the girl's mother could take her back. All are heroes who chose to risk their lives and love others in the face of unspeakable cruelty and evil.

"Too often, especially in times of change, especially in times of anxiety and uncertainty, we are too willing to give in to a base desire to find someone else, someone different, to blame for our struggles," President Obama said at the ceremony: "So here tonight we must confront the reality that around the world anti-Semitism is on the rise. We cannot deny it."

Afterward, President Obama could not have been warmer; he patted me on the shoulder, holding a long gaze as he remarked about Dad's extraordinary story. He chatted and clasped hands with Regina. And as we finished talking, President Obama showed his lighter side.

"Chris, when you finished your remarks, I leaned over to Steven [Spielberg] and said, 'I think there's a movie here.'"

"Yes, Mr. President," I said. "And guess who's in it? *You* are."

"Great," he said.

"Who would you like to play your part?" I asked.

"Me?" he said, grinning.

I returned the smile.

"Sir, I'll see if I can get you the part."

As the president moved down the receiving line, Senator Lamar Alexander tried to introduce him to "Mr. Tanner," but President Obama cut him short: "I *know*—don't mess with Lester!"

The president was clearly impressed by the vigor of these sturdy World War II vets like Lester and Paul and Sonny. "I want to see what these guys are eating, though!" he said, to loud laughter. "This is what I want to look like when I'm their age."

I hope that the next remarkable ceremony to honor Dad will happen if he is awarded the Medal of Honor or the Congressional Gold Medal.

I think all of us have been honored to know Dad's story and to have met just a few of the hundreds of thousands of noble men we call the Greatest Generation. Those young men—many of them still boys just out of high school—marched into battle with a profound sense of patriotism and duty. All faced the unimaginable horrors of war. All fought with grit, honor, sacrifice, and dignity. They secured for us a world of freedom, a world free of the terrors of Nazism.

I've had the honor to meet World War II veterans who served with my father, from Fort Jackson to Camp Atterbury, from the Battle of the Bulge to their unspeakably brutal experiences as prisoners of war.

To this day, I remain in awe of them.

Some, like my father and many of his fellow GIs, have been promoted to their post in the heavenlies. Taps played for them for the last time as reveille joined the refrain.

Others, like Lester, Sonny, Hank, and more, remain on duty here below. May we love and cherish them and all veterans who have served and will ever serve our noble nation.

Since embarking on this journey, I've gone to places I never expected to visit, been introduced to people I never expected to meet; it's helped me learn truths about my father and humanity I never imagined. Each of us, I've come to understand, has the moral capacity to make a difference in the lives of others. The good we do creates a lasting legacy that carries on from generation to generation. We don't need to do a grand thing, a huge act of courage; it's the everyday acts of goodwill that help make our world a better place. What we often don't realize is that we're leaving a legacy every time we do something kind, generous, or thoughtful for someone else—simply because, as Lester said, it's "the right thing to do."

Like Dad, we must choose right, oppose wrong, and dignify life. If we're going to make our world better, we must aspire to the place where God resides and put others ahead of ourselves.

It's that simple.

I hope Dad's story serves as a clarion call to do what's right and look to the needs of others.

When I'm speaking at schools, I often tell children and teenagers, "Listen, when you wake up in the morning, think of others first and yourselves last."

The way I see things, it would solve a lot of the world's problems if we all did that.

That's how Dad lived.

While starving and suffering alongside his men in the Nazi POW camp, Dad heard the calling from God to sing gospel songs. Dad had never sung publicly. But he made a vow to God that he would sing, if the Lord would save him from that hellhole.

God did save him, and Dad came home singing.

And he never stopped. That's how he met my mom, in fact—in church, singing, when my grandmother became his pianist. And that's how my brother and I showed up.

That's why I've had the honor to share Dad's story with you.

I pray your journey will be filled with the best of life and love guided by the good Lord above.

Acknowledgments

With profound gratitude, I acknowledge the following family and friends, old and new, who helped me along the way. Without them, Dad's story would have remained unknown. All of them are ordinary people making an extraordinary difference in our world.

To my father, Roddie: Thank you for being my hero before I knew of your heroism. Your ability to enjoy life with extraordinary moral clarity inspires us all. Thank you for showing us fearless courage, holy humility, unwavering humanity, and eternal joy. Earth and heaven are richer because of you.

To my mom, Mary Ann: Thank you for your unconditional love, your tireless work ethic, your attention to meeting needs, and for the thousands of times you cooked great food for us and picked up after us. Most of all, thank you for loving Dad and us well.

To my wife and high school sweetheart, Regina: Thank you for your uncommon innocence, undying love, unquestionable faith, and unceasing service. Your gift and energy to help our family and anyone you meet is remarkable. Thank you for your insightful suggestions along the way. You are my joy on my journey.

To my lovely daughters and families—Alicia, Steve, Austin, Haylee, and Lilly; Kristen, Jed, Hagen, Holden, and Ruthie; Lauren, Ray, Maylee, Mollee, and Roderick: Thank you for loving God and for giving us the coolest grandkids on the planet. Without you, our incredible discovery would not have happened.

Thank you, Lauren, for your interest in your papaw that sparked our journey. Your inspiring project helped me find out more about Dad—and life—than I ever dreamed.

Thank you, Alicia, for your tireless support in typing transcripts, organizing files, editing manuscripts, tending social media, sending emails, and a host of other vital tasks.

Thank you, Kristen, for your middle school project on Papaw Roddie that providentially prepared us for what was to come. Papaw Roddie would be so proud of his "three peas in a pod."

Thank you, Austin, for being a ready companion in some of our travels. Your friendship with Lester and others has been an added bonus. Thank you for enjoying our journey and new friends as much as us.

To my brother, Mike, who saved me from drowning when I was five and keeps beating his forty-year battle with schizophrenia: Keep fighting the tough fight and helping Mom.

To my sister, Priscilla Davenport; her husband, Paul; and family: Meeting you has been one of my greatest joys. Thank you for sharing your unconditional love.

To my in-laws, Glenn and Doris Grubb, who loved me like their own and left a legacy of secret generosity wherever they went: Your friendship with Mom and Dad was a blessing.

To my friend since seventh grade and brother-in-law Rodney Grubb and family: Muffett, Mikel, Tyler, Dottie, Nick, and Elisha. Thank you for cheering for Dad and his heroic "boys." Your love and tireless support inspires me to keep sharing the story. Rodney, when I first shared the story with you, you said, "That's so like your dad." You knew him like I did, and, like me, you weren't surprised by his heroics. Thank you for helping me start Roddie's Code and helping get the story out. I couldn't have done it without you.

To my longtime friend Rick Oster who also helped me start Roddie's Code: Thank you for your friendship, support, and for your love for dad and his story. You knew Dad too. Once you heard the story, what you knew about him fit perfectly with his character. No surprise. And you too wanted to help tell the story far and wide. Thank you for helping do that.

Thank you to my good friend attorney Jim Tipton of Gentry, Tipton & McLemore. Your favor extended toward me and Dad in sharing legal guidance has been instrumental in helping organize Roddie's Code, review contracts, define legal processes, and even telling me where I can find the best sandwich downtown. Thank you for your ready help. Most of all, thank you for your friendship.

Thank you to another good friend, Chuck Morris of Morris Creative Group, for favoring Dad and Roddie's Code with services beyond our means. The logo you designed for Roddie's Code is perfect—clean, impactful, and easily understood. Thank you and your staff for helping us with all things marketing.

To my sister-in-law Robbyn Collins and family: Erin, Drew, Brittney, Bradyn, Kaydence, and Blayklee. Thank you for your continued love for Dad. Your excitement about his story is an encouragement.

To Lester Tanner and family: Thank you for sharing Dad's story and inviting us to join your family's journey—a family we cherish. Lester, you've touched us with your sincerity, wisdom, warmth, and generous spirit. Your righteous character and love for life moves us to do what's right for others. Your continued friendship with Dad and me has blessed my family and the world. Thank you for helping me start Roddie's Code. Without your advice and leadership I would have never done it. Thank you being one of Dad's best men, then and now.

To Paul and Corinne Stern and family: Thank you for sharing your home, your life, your amazing family and most of all your love. Paul and Corinne, your romantic story is epic and your radiant lives are exemplary. Paul, thank you for warming our hearts with your infectious smile and natural cheerfulness. Your chilling stories of serving as a combat medic made us gasp, and your heroic journey through life made us cheer. Thank you for showing all of us the power of family and love.

To Sydney "Skip" Friedman and family: Thank you for your kindness and love. We loved meeting your extended family and our visits always felt like we were home. Thank you, Skip, for being a terrific friend, husband, father, grandfather, citizen, leader, and person. Your zest for life and your zeal for goodness have made all of us better. I wish I could have known your precious Penny. When you spoke of her your eyes sparkled with life.

To Irwin "Sonny" Fox and Cely: Thank you for showing Regina and me what it means to love deeply and have fun while doing so. Sonny, you are one of a kind and loved by many. Everywhere I share Dad's story I meet folks in their sixties and seventies known as "Sonny's kids" from watching *Wonderama*. All of them glow with fond memories of you. Regina and I do, too.

To Henry "Hank" Freedman: Thank you for your warmth and friendship. Becoming friends with you has been one of the highlights of my journey. When we met I felt an instant connection. Little did I know it was our shared faith. Thank you for sharing your Christian journey and special moments of Dad's faith during your horrific war experiences. Our family will forever cherish you and your vibrant witness of God's grace and goodness.

To Jack and Marcia Lou Sherman: Thank you, Mrs. Sherman, for helping Jack and me connect. How fondly I remember Jack's ready wit and winsome spirit. I'm sure Jack and Dad were fast friends during the war. Thank you, Jack, for sharing details about Dad I could never have known. I only wish we could have met in person.

To Lorna Cerenzia Nocero: Thank you for calling back a stranger from Tennessee and then taking time to visit the city to meet with Lester and me. At Camp Atterbury, Dad, Frankie, and Lester were inseparable. It seemed to me that they were back together that day at the Harvard Club. I am forever grateful for their friendship that helped sustain them during a horrific time in their lives. Thank you for sharing stories about your wonderful dad. Frankie was Dad's best friend from basic training through liberation. Your dad's love for life and kindness to others was a blessing to my father. Lester and I are grateful that their blessed friendship continues today through all of us.

To my friends at 3 Arts Entertainment: Erwin Stoff and Richard Abate. Thank you both for loving Dad's story and taking the risk to invest in me. Thank you, Erwin, for our friendship and for always taking time for me. I enjoy our insightful conversations, and I look forward to many more.

Thank you, Richard, for tracking me down, following providential direction from a publishing colleague. Thank you for believing in me and wanting the world to hear Dad's story. I'm so glad we are friends. Your expertise, ready encouragement, straightforward leadership, and wonderful staff have made all the difference. Thank you for finding the perfect match with my coauthor and my publisher. You are the boss and a rock star.

To my coauthor, Douglas Century: What can I say? You still like me after all the bad phrasing, overblown adjectives, incomplete sentences, southernisms, and endless edits. I'm blessed to call you my coauthor. When the Lord teamed us up He knew what He was doing. I know Richard gets some of the credit, but I believe the Lord was working behind the scenes to blend two very different people into perfect harmony. Thank you for being a fantastic friend and even better writer, for teaching me the power of words, and for connecting Dad's story with Stéphane Guevremont, military historian in Canada, who helped us find key information on the German officer, Major Siegmann. Most of all thank you for loving Dad's story and wanting to tell it as much as I do.

To my friends at HarperOne—Judith Curr, Miles Doyle, Suzanne Quist, and all of the amazing folks at Harper. Judith, you've made Regina and me feel like family—a family we cherish. Thank you for choosing to share Dad and his brave men with the world. Your contribution to Dad's legacy will long endure, and for that we are grateful. We are also grateful for the terrific support you and your team have given.

Miles, thank you for leading that fantastic support. Your instincts and suggestions as editor have been spot-on. Every cut and addition, every probing question that made me wrestle with my thoughts, have paid off. And your uncanny ability to tell me "no" in a nice way has been masterful. You are really good at this. Thank you for your friendship and for helping Dad's powerful story come alive in us.

To the wonderful congregations of West Haven Baptist Church in Knoxville, Tennessee, and Piney Grove Baptist Church in Maryville, Tennessee (and every church who put up with me): Thank you for your endless love and ceaseless mercy. Your support has been heaven-sent.

To YOKE Youth Ministries, John and Helen Coatney, the staff, board, and everyone in our YOKE family: My journey of discovery began in 2009 while serving middle school kids with you. Thank you for your unconditional love and boundless support. Thank you for continuing to love middle school children. Never stop. You are making a big difference in their lives.

Thank you to my former and current Tennessee state and national leaders and their staffs: Senator Lamar Alexander, Senator Bob Corker, Senator Marsha Blackburn, Congressman John Duncan Jr., Congressman Tim Burchett, Congressman Phil Roe, Congressman Chuck Fleischmann, Congressman Scott DesJarlais, Congressman Jim Cooper, Congressman John Rose, Congressman Mark Green, Congressman David Kustoff, Congressman Steve Cohen, Governor Bill Haslam, Governor Bill Lee, State Representative Bob Ramsey, and State Representative Jerome Moon.

Special thanks to Jenny Stansberry, Heather Hatcher, Rhonda

Smithson, and Chris Coyne. Thank you for helping me pursue the Congressional Medal of Honor for Dad by putting together a comprehensive recommendation package. Thank you for your hard work over more than two years. While we are still pursuing the Medal of Honor for Dad, which I believe he will receive some day, your excellent work has led to other high honors and has helped propel this book into being. It has been an honor to serve alongside you.

To our dear friends, Larry and Barbara Goldstein: Thank you for loving us well and for sharing Dad's story with Yad Vashem. Your personal touch made all the difference in the world. Regina and I had no idea what you were up to when you kept asking for more information, but we were thrilled when we found out. We are still thrilled. Most of all we rejoice to have such wonderful friends like you.

To my friends at Yad Vashem—Avner Shalev, Rabbi Yisrael Meir Lau, Leonard Wilf, Ron Meyer, Dorit Novak, Irena Steinfeldt, Simmy Allen, Shaya Ben-Yehuda, Searle Brajtman, Dr. Susanna Kokkonen, Sari Granitza, Shavit Shimon, Malka Weisberg, Ephraim and Stephanie McMahon Kaye, and all at Yad Vashem who have touched my life with goodness: Thank you and the nation of Israel for our treasured friendship and for bestowing your highest honor on my father. Our family is forever grateful.

To our wonderful friend Stanlee Stahl and the Jewish Foundation for the Righteous: Thank you, Stanlee, for calling a perfect stranger in August 2016 and asking me if I would like to go with you and a film crew to follow my father's footsteps through Belgium and Germany. Without hesitation I said, "Of course I'd like to go." And we went. While in Germany, we shot enough film for two award-winning films thanks to our excellent production crew of Paul Allman, Dean Beals, Petr Cikhart, and others. And little did you know, Stanlee, that you were answering one of my prayers. In late 2014 I began praying for a way to follow my father's footsteps during WWII. I wanted to go but didn't have the money. So the Lord sent you. And all it cost me was my time and lots of energy. Thank you for following your heart. You were more than an answer to my prayers. You and JFR have become great friends and enablers in helping extend Dad's legacy. Thank you for honoring Dad with the Yehi Or award. Dad truly is a light to the world. Our family cherishes our deep friendship and we are eternally grateful for your favor toward Dad.

Thank you to Carl Wouters and Doug Mitchell for your expertise in all things WWII and the 106th Infantry. Without your knowledge and guidance, Dad's book would not be complete and our film crew would still be lost somewhere in the Ardennes. Carl, I was amazed when you took me to the very spot of Dad's capture and described the chilling scene. You did the same when we slipped down to the train station at Bad Orb a few days later around midnight. I will never forget your description of our American doughboys being marched through those narrow, angry streets on Christmas Day 1944. Thank you and Doug for your friendship and for a job well done.

Thank you to my friend Penny Simon for connecting me with my new friend Ted Koppel, who did narration on our latest film. Penny, you are awesome. Your advice on publicity has been invaluable. Thank you, Ted, for lending your professionalism and excellence to Dad's story. Our family is grateful for your kindness and generosity. As one of your longtime fans, I think it is so cool to have you in the film. Thank you, sir.

To Arlene and Paul Samuels: Thank you and AIPAC for taking a chance on me and inviting me to speak about Dad for the first time for Rabbi Avi Perets at Temple Emanu-El in Myrtle Beach, South Carolina. The congregation couldn't have been any friendlier and loving. They were great and so was the media coverage.

Regina and I enjoyed our time with you and Paul. Thanks for your friendship and for helping put the wheels in motion for sharing Dad's story with the world then and now.

Thank you to my friends Howard Kohr, Robert Cohen, Brad Gordon, and everyone at AIPAC for allowing me to speak in support of our beloved Israel. Thank you for helping keep Israel and America strong and united. Thank you for your support of Dad's Congressional Gold Medal bills.

To my friends Pastor John and Diana Hagee, Shari Dollinger, Lyndon Allen, and others at CUFI: Thank you for standing with Israel and allowing me to be a small part of your vital work. Your heaven-sent ministry and enduring love for the Holy Land is making Israel stronger and safer. Thank you for your support of Dad's Congressional Gold Medal bills.

To Jonathan Greenblatt and the Anti-Defamation League: Thank you for your friendship and for your support of Dad and his legacy. Our family is grateful for your help with Congress regarding Dad's Gold Medal bills.

To Danise Peters and Cathy Hinesley, better known as Thelma and Louise: Thank you for giving me a guided tour of Israel as we bolted across the country without directions heading toward Haifa. By God's grace we made it there and back and have stories for our grandchildren. Thank you for loving God, loving Israel, loving Dad, and putting up with me.

To Dr. Jerry Westbrook, Jim Snyder, Rachel McRae, Preston Trotter, my church family, and a host of others too numerous to mention: Thank you for your faithful support of Dad and never getting tired of hearing about him or his men.

Most important, to all US veterans, the 106th Infantry "Golden Lions" Division, the 422nd Regiment, every American POW at Stalag IXA, and especially the 255 GIs who are listed in and signed Dad's journal: We salute all of you.

Notes

1   *"Look back and smile . . .":* Sir Walter Scott, *Kenilworth*, 1821.

37   *"Leadership is intangible . . .":* General Omar Bradley, speaking at Command and General Staff College, May 16, 1967.

44   *"I am a little boy . . .":* Roderick Edmonds, Letters to Santa, "Roderick Edmonds, Thinks of Crippled Boy," *The Knoxville News-Sentinel*, December 13, 1926, p. 5.

46   *"The hope that had almost . . .":* Ray Hill, "The Great Depression in Tennessee," *The Knoxville Focus*, January 19, 2016, http://knoxfocus.com/archives/great -depression-tennessee/.

50   *"You are now on your way . . .":* Author's artifacts (Roddie Edmonds's original induction letter).

56   *"The greatest attack that has ever been . . .":* President Franklin D. Roosevelt, as quoted by UPI, "FDR Urges 'Total Defense,' Warning Nation Is Menaced by Foes Without and Within," UPI Archives, September 2, 1940, https:// www.upi.com/Archives/1940/09/02/FDR-urges-total-defense-warning -nation-is-menaced-by-foes-without-and-within/5341504232633.

61   *"Winston Churchill saw a spectacular . . .":* *The State*, Columbia, South Carolina, December 22, 2017.

61   *"They're just like money in the bank":* Winston Churchill quoted in *The State*, Columbia, South Carolina, December 22, 2017.

64   *"I do not know beneath what sky . . .":* Richard Hovey, "Unmanifest Destiny" in *Along the Trail* (New York: Duffield, 1907), 16.

64   *"jet-black hair . . .":* Martin King, Ken Johnson, and Michael Collins, *Warriors of the 106th* (Philadelphia: Casemate, 2017), Kindle edition.

67   *"I was six years old in 1929 . . .":* Lester Tanner, interview with author, New York, NY, March 20, 2013, and April 17, 2017.

74   *"I know you are all waiting . . .":* Paul Osmundson, "Flashback Friday: Winston Churchill Visits Fort Jackson in 1942," *The State*, December 22, 2017, https://www.thestate.com/news/local/article191204024.html.

76   *"We took basic infantry training . . .":* Sydney "Skip" Friedman, interview with author, Shaker Heights, Ohio, September 19, 2014.

76 *"largest educational program . . ."*: Louis E. Keefer, *Scholars in Foxholes: A Study of the Army Specialized Training Program in World War II* (Jefferson, NC: McFarland, 1988).

77 *"soldiers first, students second"*: Keefer, *Scholars in Foxholes*, 43.

78 *"gambit designed to keep . . ."*: "Wasting the Best and the Brightest: The ASTP Program in World War Two," Daily Mercury Chronicle, November 16, 2014, https://dailymercurychronicle.wordpress.com/2014/11/18/967/.

79 *"democratize American society . . ."*: Louis E. Keefer, "The Army Specialized Training Program in World War II," accessed July 26, 2019, http://www.pierce-evans.org/ASTPinWWII.htm.

81 *"the withdrawals were often . . ."*: Robert Roswell Palmer, Bell Irvin Wiley, and William R. Keast, *United States Army in World War 2, Army Ground Forces, Procurement and Training of Ground Combat Troops* (Center of Military History, 1948), 472.

81 *"And we're sending men to college?"*: Phillip Leveque, "ASTP: The Army's Waste of Manpower," http://www.89infdivww2.org/memories/levequeastp.htm.

82 *"What kind of soldiers . . ."*: Keefer, "The Army Specialized Training Program in World War II."

89 *"Our job was to help . . ."*: Paul Stern, interview with author, Reston, Virginia, November 1, 2013.

92 *"It was one of the most remarkable . . ."*: Antony Beevor, *Ardennes 1944: The Battle of the Bulge* (New York: Viking, 2015), 4.

93 *"militarily, the war is over"*: Stephen W. Sears, *Eyewitness to History: World War II* (New Word City, 2015), Kindle edition.

93 *"Victory is everywhere . . ."*: Sears, *Eyewitness to History*, Kindle edition.

94 *"Damn, the war will be over . . ."*: Sonny Fox, *But You Made the Front Page! Wonderama, War, and a Whole Bunch of Life* (Argo-Navis, 2012), Kindle edition.

95 *"Dearest Folks, I'm sitting . . ."*: Fox, *But You Made the Front Page!*, Kindle edition.

97 *"worse than fighting in . . ."*: Beevor, *Ardennes 1944*, 76.

97 *"The Hürtgen Forest sat along . . ."*: Fox, *But You Made the Front Page!*, Kindle edition.

97 *"As we were walking up . . ."*: Fox, *But You Made the Front Page!*, Kindle edition.

98 *"In the Hürtgen Forest . . . the way the shells . . ."*: Stern, interview with author.

100 *"brought to the camp . . ."*: Konstanin Simonov's 1944 reporting for *Krasnaia Zvezda*, quoted in Joshua Rubenstein, "What They Saw: The Liberation of Majdanek 70 Years Ago Today," Cognoscenti, July 23, 2014, https://www.wbur.org/cognoscenti/2014/07/23/nazi-death-camps-world-war-ii-joshua-rubenstein.

100 *"I have just seen . . ."*: W. H. Lawrence, "Nazi Mass Killing Laid Bare in Camp," *New York Times*, August 30, 1944.

101 *"Be proud of your assignment . . .":* King, Johnson, and Collins, *Warriors of the 106th*, Kindle edition.

101 *"other much-appreciated goodies":* John W. Morse, *The Sitting Duck Division* (San Jose, CA: Writers' Club Press; iUniverse, 2001), Kindle edition.

101 *"Our ship was large . . .":* Morse, *Sitting Duck Division*, Kindle edition.

105 *"I can remember thinking . . .":* Ernest Hemingway, *Across the River and into the Trees* (New York: Charles Scribner's Sons, 1950).

108 *"I shall go over to the offensive . . .":* Adolf Hitler quoted in Hugh M. Cole, *The Ardennes: Battle of the Bulge*, United States Army in World War II, The European Theater of Operations (Washington, DC: Center of Military History United States Army, 1965), 2.

109 *"It was obvious to me that . . .":* Karl Gerd von Rundstedt quoted in Robin Cross, "The Battle of the Bulge," BBC News, last updated February 17, 2011, https://www.bbc.co.uk/history/worldwars/wwtwo/battle_bulge_01.shtml.

109 *"All Hitler wants me to do is . . .":* Patrick Delaforce, *The Battle of the Bulge: Hitler's Final Gamble* (South Yorkshire, England: Pen and Sword, 2004), 7.

109 *Hitler didn't think the US army . . . :* Peter Caddick-Adams quoted in Simon Worrall, "Book Talk: The Real Reason Hitler Launched the Battle of the Bulge," *National Geographic* online, December 15, 2014, https://news.nationalgeographic.com/news/2014/12/141214-battle-of-the-bulge-hitler-churchill-history-culture-ngbooktalk/.

111 *"Full field packs, helmet, rifle . . .":* Morse, *Sitting Duck Division*, Kindle edition.

111 *Though he would be heading . . . :* Tanner, interview with the author.

112 *"old-growth pine-forest . . .":* Robert M. Citino, "First Blood on the Ghost Front," *World War II*, December 2014, https://www.historynet.com/first-blood-ghost-front.htm.

112 *"vacant landscape of trackless woods . . .":* Citino, "First Blood on the Ghost Front."

112 *According to an after-action report . . . :* William P. Moon, "Operations of the 1st Battalion, 422D (106th Infantry Division) in the Battle of the Bulge, in the Vicinity of Schlausenbach, Germany, 10–19 December 1944 (Ardennes—Alsaac Campaign)," Advanced Infantry Officers Course, 1949–1950, personal experiences of Major William P. Moon (Fort Benning, GA: USAIS Library), 5.

113 *"Lucky guys!":* John Toland, *Battle: The Story of the Bulge* (New York: Random House, 1959; New York: Bison Books, 2016), Kindle edition.

114 *"New arrivals needed a chance . . .":* John C. McManus, *Alamo in the Ardennes: The Untold Story of the American Soldiers Who Made the Defense of Bastogne Possible* (Hoboken, NJ: John Wiley & Sons, 2007), 7.

114 *"The disposition of troops . . .":* Moon, "Operations of the 1st Battalion, 422D," 4.

115 *"in poor condition . . .":* Moon, "Operations of the 1st Battalion, 422D," 5.

115 *"There were mine fields and . . .":* Moon, "Operations of the 1st Battalion, 422D," 6.

115 *"We were among the first troops . . .":* Friedman, interview with author.

116 *"Both the enemy and the weather . . .":* Rick Atkinson with Kate Waters, *Battle of the Bulge,* Young Readers Adaptation (New York: Henry Holt, 2015), Kindle edition.

116 *"assumed that the engine noises . . .":* Beevor, *Ardennes 1944,* 109.

119 *"made up of seven buildings . . .":* Irwin J. Kappes, "Hitler's Ultra-Secret Adlerhorst," Military History Online, March 15, 2003, https://www .militaryhistoryonline.com/wwii/articles/adlerhorst.aspx.

121 *"This battle is to decide . . .":* Alex Kershaw, *The Longest Winter: The Battle of the Bulge and the Epic Story of World War II's Most Decorated Platoon* (Cambridge, MA: Da Capo Press, 2005), Kindle edition.

122 *"conceited, reckless military leader . . .":* Beevor, *Ardennes 1944,* 88.

124 *"almost worshipped Skorzeny . . .":* Beevor, *Ardennes 1944,* 93.

124 *"Skorzeny, this next assignment . . .":* Beevor, *Ardennes 1944,* 92.

124 *"I want you to command . . .":* Rupert Butler, *The Black Angels: A History of the Waffen-SS* (New York: St. Martin's Press, 1979), 183–84.

125 *"Officers and NCOs . . .":* Beevor, *Ardennes 1944,* 92.

125 *"how to tap their cigarette . . .":* Beevor, *Ardennes 1944,* 93.

126 *"It is plain that . . .":* Leo Barron, *Patton at the Battle of the Bulge* (New York: Dutton Caliber, 2015).

126 *A civilian named Elise Dele . . . :* Andrew Knighton, "Allies Knew About Battle of the Bulge—Why They Ignored Intelligence," War History Online, https://www.warhistoryonline.com/world-war-ii/warning-battle-of-the -bulge.html.

127 *"It's the Ardennes!":* John C. McManus, *Alamo in the Ardennes: The Untold Story of the American Soldiers Who Made the Defense of Bastogne Possible* (New York: Dutton Caliber, 2008).

127 *"Word of the forthcoming offensive . . .":* Beevor, *Ardennes 1944,* 102.

127 *"vital information does not . . .":* Beevor, *Ardennes 1944,* 102.

127 *"No goddamn fool . . .":* Sid Moody, "The Battle of the Bulge: Victory on the Ghost Front," Associated Press, December 14, 1994.

127 *"needed to husband their strength . . .":* Beevor, *Ardennes 1944,* 103.

128 *"but the Allied command had gravely . . .":* Beevor, *Ardennes 1944,* 103.

131 *"Our little village of Schlausenbach . . .":* Lester Tannenbaum, "Eleven Days in December: December 10–21 in Germany on the Schnee Eifel, Battle of the Bulge" (unpublished manuscript, August–September 1945), hard copy.

131 *"bored through the darkness . . .":* Atkinson with Waters, *Battle of the Bulge,* Kindle edition.

132 *"Strange lights lit up the sky . . .":* Morse, *Sitting Duck Division,* Kindle edition.

132 *"idly chatting in front . . .":* Morse, *Sitting Duck Division,* Kindle edition.

132 *"piercing the misty gloom . . .":* Morse, *Sitting Duck Division,* Kindle edition.

133 *"like wolves on a sheepfold . . ."*: Atkinson with Waters, *Battle of the Bulge*, Kindle edition.

133 *"On no segment of . . ."*: Rick Atkinson, *The Guns at Last Light: The War in Western Europe, 1944–1945*, vol. 3, Liberation Trilogy (New York: Henry Holt, 2013), 432.

134 *"How long has this been going on . . ."*: Fox, *But You Made the Front Page!*, Kindle edition.

136 *"But I'm General Bruce Clarke . . ."*: Toland, *Battle: The Story of the Bulge*, Kindle edition.

136 *"Three times I was ordered . . ."*: Omar Nelson Bradley, *A Soldier's Story* (New York: Henry Holt, 1951), 467.

137 *"A half million GIs played . . ."*: Bradley, *A Soldier's Story*, 185.

138 *"You know how things are . . ."*: Toland, *Battle: The Story of the Bulge*, Kindle edition.

139 *"It was a case of every dog . . ."*: Atkinson with Waters, *Battle of the Bulge*, Kindle edition.

139 *"the fear-crazed occupants . . ."*: Atkinson with Waters, *Battle of the Bulge*, Kindle edition.

139 *"Take a ten-minute break . . ."*: Atkinson with Waters, *Battle of the Bulge*, Kindle edition.

140 *"There are arrows north . . ."*: Friedman, interview with author.

141 *"nine thousand GIs . . ."*: Atkinson with Waters, *Battle of the Bulge*, Kindle edition.

141 *"None of us knew how serious . . ."*: Tanner, interview with author.

143 *The eleven men tried . . .* : For more on the Wereth 11 massacre, see Denise George and Robert Child, *The Lost Eleven: The Forgotten Story of Black American Soldiers Brutally Massacred in World War II* (New York: Dutton Caliber, 2017).

144 *"to seek shelter or run . . ."*: Beevor, *Ardennes 1944*, 145.

144 *"rings, cigarettes, watches . . ."*: Beevor, *Ardennes 1944*, 145.

146 *"herded together into . . ."*: Beevor, *Ardennes 1944*, 147.

146 *"Immediate publicity . . ."*: Beevor, *Ardennes 1944*, 147.

146 *"It is dangerous at any . . ."*: Beevor, *Ardennes 1944*, 162.

146 *"What utter madness to . . ."*: Beevor, *Ardennes 1944*, 147.

149 *"We hadn't taken many casualties . . ."*: Fox, *But You Made the Front Page!*, Kindle edition.

150 *"one of the guys in . . ."*: Fox, *But You Made the Front Page!*, Kindle edition.

153 *"I was the only one who spoke German . . ."*: Stern, interview with author.

154 *it was a "casualty assignment" . . .* : Tannenbaum, "Eleven Days in December."

154 *"Of all the guys . . ."*: Tannenbaum, "Eleven Days in December."

155 *"The fellas know we're zipped . . ."*: Friedman, interview with author.

160 *"Showers, warm beds . . ."*: Colonel R. Earnest Dupuy, *St Vith: Lion in the Way: The 106th Infantry Division in World War II* (Washington, DC: Infantry Journal Press, 1949), 148.

160 *"We're still sitting like . . ."*: Atkinson, *The Guns at Last Light*, 437.

163 *"One of the things . . ."*: Tanner, interview with author.

168 *"We were lined up . . ."*: Morse, *Sitting Duck Division*, Kindle edition.

168 *"suffered the most . . ."*: Morse, *Sitting Duck Division*, Kindle edition.

168 *"If you didn't march . . ."*: Friedman, interview with author.

169 *"As our march into Germany . . ."*: Morse, *Sitting Duck Division*, Kindle edition.

169 *"A couple of fat German officers . . ."*: Richard Peterson, *Healing the Child Warrior* (Largo, FL: CombatVets Network, 1992, 2000), 24.

169 *"There were distinct differences . . ."*: Friedman, interview with author.

170 *"We heard we were being . . ."*: Friedman, interview with author.

170 *"We would remain as captives . . ."*: Peterson, *Healing the Child Warrior*, 25.

171 *"The fear of God . . ."*: Rev. A. C. Dixon, D. D., "The Hero of Faith," in *Northfield Echos*, vol. 2., ed. D. L. Pierson (East Northfield, MA: Rastall and McKinley, 1895), 20.

174 *"The boxcars turned out to be . . ."*: Sonny Fox, interviewed in *Footsteps of My Father*, directed by Paul Allman (West Orange, NJ: Jewish Foundation for the Righteous, 2018), documentary.

174 *"rendering thoughts of escape hopeless"*: Fox, *But You Made the Front Page!*, Kindle edition.

174 *"We couldn't sit down . . ."*: Tanner, interview with author.

175 *"We traded insults . . ."*: Peterson, *Healing the Child Warrior*, 11.

175 *"claustrophobic hell"*: King, Johnson, and Collins, *Warriors of the 106th*, Kindle edition.

177 *"The earth shook with explosions . . ."*: Morse, *Sitting Duck Division*, Kindle edition.

177 *"The British were bombing the rail yard . . ."*: Fox, interviewed in *Footsteps of My Father*.

177 *"From the other side of the boxcar . . ."*: Henry "Hank" Freedman, interview with author, Suwanee, Georgia, 2014.

182 *"The words were different . . ."*: Morse, *Sitting Duck Division*, Kindle edition.

183 *"deliberate attempt at dehumanization"*: Peterson, *Healing the Child Warrior*, 28.

184 *"Going through the lovely . . ."*: Friedman, interview with author.

184 *"built-in reserves . . ."*: Peterson, *Healing the Child Warrior*, 30.

184 *"to find a crystalline . . ."*: Peterson, *Healing the Child Warrior*, 30.

186 *"The Nazis' goal . . ."*: Friedman, interview with author.

188 *"One of the first things they did . . ."*: Freedman, interview with author.

188 *"An American officer who happened . . ."*: Freedman, interview with author.

189 *"Very few of the Jewish prisoners . . ."*: Tanner, interview with author.

189 *"I was politically aware enough . . ."*: Ernest Kinoy interviewed in *Berga: Soldiers of Another War*, written and directed by Charles Guggenheim (Arlington, VA: Public Broadcasting System, 2002), documentary.

189 *Sonny Fox recalled getting . . .* : Fox, *But You Made the Front Page!*, Kindle edition.

191 *"filthy pads filled with dried . . ."*: Peterson, *Healing the Child Warrior*, 32.

191 *it was part of the Germans' . . .* : Friedman, interview with author.

191 *"It isn't what happens . . ."*: Earl F. Verham, unpublished letter written from Stalag IXA, April 1, 1945, http://www.pegasusarchive.org/pow/earl _verham.htm.

191 *"'I want you to do two things . . .'"*: Fox, *But You Made the Front Page!*, Kindle edition.

192 *"Surreptitious looks at comrades . . ."*: Peterson, *Healing the Child Warrior*, 45.

193 *"The only good thing one can say . . ."*: Sonny Fox, interview with the author, New York, NY, September 15, 2015.

194 *"a paving block than anything . . ."*: Ernest Kinoy, "A Walk Down the Hill," CBS, March 18, 1957, teleplay.

194 *"That small loaf of what . . ."*: Tanner, interview with author.

194 *"Some way had to be . . ."*: Morse, *Sitting Duck Division*, Kindle edition.

195 *"The bread started out being . . ."*: Fox, *But You Made the Front Page!*, Kindle edition.

195 *"A friend and I got into . . ."*: Peterson, *Healing the Child Warrior*, 49.

196 *"Faith was something . . ."*: Russell Gunvalson, interviewed by Thomas Saylor for the POW Oral History Project, February 29, 2004.

196 *"ubiquitous and the size of . . ."*: King, Johnson, and Collins, *Warriors of the 106th*, Kindle edition.

197 *"Most of their supplies" were "first aid kits . . ."*: War Crimes Investigation, 1945 (NARA; Washington, DC: National Archives).

198 *"We protested that we . . ."*: Tanner, interview with author.

199 *Belgian and French military records . . .* : E. Gillet, "Histoire des Sous-Officiers et Soldats Belges Prisonniers de Guerre, 1940–1945 (Fin)," *Revue belge d'histoire militaire* 28, no. 5 (March 1990): 350–83.

199 *The Oberkommando der Wehrmacht . . .* : Georges Baud, Louis Devaux, and Jean Poigny, "Mémoire Complémentaire sur Quelques Aspects des Activités du Service Diplomatique des Prisonniers de Guerre: S.D.P.G.—D.F.B.— Mission Scapini, 1940–1945" (unpublished paper; Paris, France: January 1984), hard copy.

199 *This little-known order*: OKW Order Az 2 f. 24 73 o Kriegsgef. Allg Nr. 2140/42 Kennzeichnen der Juden—"Marking the Jews," March 3, 1942.

199 *Jewish soldiers would not . . . :* Baud, Devaux, and Poigny, "Mémoire Complémentaire sur Quelques Aspects."

200 *"We as the delegates always protested . . .":* Pierre Arnaud, full testimony, in Baud, Devaux, and Poigny, "Mémoire Complémentaire sur Quelques Aspects."

200 *"I personally visited many such . . .":* Arnaud, in Baud, Devaux, and Poigny, "Mémoire Complémentaire sur Quelques Aspects."

200 *"we were dealing with officers . . .":* Arnaud, in Baud, Devaux, and Poigny, "Mémoire Complémentaire sur Quelques Aspects."

201 *"when Siegmann saw that war . . .":* Arnaud, in Baud, Devaux, and Poigny, "Mémoire Complémentaire sur Quelques Aspects."

201 *Major Siegmann "was a very tough . . .":* Arnaud, in Baud, Devaux, and Poigny, "Mémoire Complémentaire sur Quelques Aspects."

202 *"Most of the American troops . . .":* Tanner, interview with author.

203 *"That's a moral crisis for . . .":* Fox, *But You Made the Front Page!*, Kindle edition.

203 *"All Jewish boys to be separated . . .":* Roger Cohen, *Soldiers and Slaves: American POWs Trapped by the Nazis' Final Gamble* (New York: Alfred A. Knopf, 2005), 82.

203 *"An American officer came in . . .":* Cohen, *Soldiers and Slaves*, 82.

204 *"I tried to say no . . .":* Cohen, *Soldiers and Slaves*, 83.

204 *"Fortunately, the Geneva Convention . . .":* Tanner, interview with author.

205 *"The Delegate of the Protecting Power . . .":* "Conditions of Stalag IXB Bad Orb," International Red Cross declassified report, Werner Buchmuller, January 24, 1945 (NARA; Washington, DC: National Archives).

205 *"special work detail":* Arnaud, in Baud, Devaux, and Poigny, "Mémoire Complémentaire sur Quelques Aspects."

205 *"full authority":* Cohen, *Soldiers and Slaves*, 9.

205 *"Jewish prisoners of war . . .":* Cohen, *Soldiers and Slaves*, 81.

206 *"Everyone trudged along . . .":* Peterson, *Healing the Child Warrior*, 40.

207 *"You're free to go":* Tanner, interview with author.

207 *"Remember this well . . .":* Tanner, interview with author.

214 *"This was for the confidence . . .":* Hans Kasten, "The Story of the War Experience of J.C.F. Kasten IV (Hans) [During] World War II," Experiencing War, Library of Congress, October 26, 2011, https://memory.loc.gov/diglib/vhp-stories/loc.natlib.afc2001001.12002/narrative?ID=pn0001.

214 *"There were eight chairs . . .":* Kasten, "The Story of the War Experience."

214 *"Kasten, we want the names . . .":* Kasten, "The Story of the War Experience."

214 *"We are all Americans . . .":* Kasten, "The Story of the War Experience."

215 *"lying and spying against . . .":* Anthony Acevedo, Voces Oral History Project, University of Texas at Austin, https://voces.lib.utexas.edu/collections/stories/anthony-acevedo.

216 *"They put needles in my fingernails . . ."*: Acevedo, Voces Oral History Project.

216 *"crammed into box cars as before . . ."*: Kasten, "The Story of the War Experience."

216 *"Berga was a slave labor . . ."*: Kasten, "The Story of the War Experience."

217 *"The agony of it all . . ."*: Kasten, "The Story of the War Experience."

217 *"seven days per week . . ."*: William J. Shapiro, "Berga am Elster: A Medic Recalls the Horrors of Berga," Jewish Virtual Library, accessed July 11, 2019, https://www.jewishvirtuallibrary.org/a-medic-recalls-the-horrors -of-berga.

218 *"Slate dust was choking . . ."*: Shapiro, "Berga am Elster."

219 *"The bodies of European . . ."*: Roger Cohen, "The Lost Soldiers of Stalag IX-B," *New York Times Magazine*, February 27, 2005.

219 *"Across the street were three . . ."*: Cohen, *Soldiers and Slaves*, 149–50.

220 *"No, I'm a Christian . . ."*: Joe Littell quoted in Cohen, *Soldiers and Slaves*, 160.

220 *"On very harsh blustery . . ."*: Shapiro, "Berga am Elster."

221 *"I had met him . . ."*: Shapiro, "Berga am Elster."

221 *"His body laid on . . ."*: Shapiro, "Berga am Elster."

224 *"a French camp . . ."*: Peterson, *Healing the Child Warrior*, 47.

229 *Suddenly, Major Siegmann approached . . .*: Lester Tanner, interview with author.

231 *"The opposite of love . . ."*: Elie Wiesel, *US News & World Report*, October 27, 1986.

233 *"We went back to the barracks . . ."*: Tanner, interview with author.

234 *"The way we were mixed . . ."*: D. B. Frampton Jr., "Ich bin ein Kriegsgefangener und Still Singing the Praises of General George S. Patton, USA," from a speech delivered to the Benjamin Franklin Chapter of Sons of the American Revolution, September 16, 1988, http://www.indianamilitary .org/German%20PW%20Camps/Prisoner%20of%20War/PW%20Camps /Stalag%20IX-A%20Ziegenhain/D%20B%20Frampton/Frampton-D-B.pdf.

235 *"For the most part . . ."*: Fox, *But You Made the Front Page!*, Kindle edition.

236 *"One man had worked . . ."*: Peterson, *Healing the Child Warrior*, 51.

236 *"You might imagine what happened . . ."*: David Dennis, "A Memory of Time Spent in a German Prison Camp," unpublished memoir, n.d.

236 *"recipes began to sound . . ."*: Fox, *But You Made the Front Page!*, Kindle edition.

236 *"We got so used to . . ."*: Stern, interview with author.

237 *"helped us build dream castles . . ."*: Frampton Jr., "Ich bin ein Kriegsgefangener."

239 *"more dead than alive"*: Peterson, *Healing the Child Warrior*, 48.

239 *"Dirt caked their faces and bodies . . ."*: Gene Kelch quoted in Peterson, *Healing the Child Warrior*, 47.

239 *"Watching the British soldiers' . . ."*: Peterson, *Healing the Child Warrior*, 48.

239 *"British prisoners arrived . . ."*: Peterson, *Healing the Child Warrior*, 48.

240 *"The French looked in good . . .":* Peterson, *Healing the Child Warrior,* 44.

240 *"They would listen to the communiqués . . .":* Fox, *But You Made the Front Page!,* Kindle edition.

241 *"The Red Army has liberated . . .":* "1945: Auschwitz Death Camp Liberated," BBC News, January 27, 1945, http://news.bbc.co.uk/onthisday/hi/dates /stories/january/27/newsid_3520000/3520986.stm.

243 *"A physiological fact about . . .":* Fox, *But You Made the Front Page!,* Kindle edition.

243 *"Captain Morgan answered medical questions . . .":* Frampton Jr., "Ich bin ein Kriegsgefangener."

245 *"You're almost ready to give up . . .":* Gunvalson, POW Oral History Project.

245 *"I thought about suicide . . .":* Peterson, *Healing the Child Warrior,* 46.

245 *"Roddie, being the communications chief . . .":* Tanner, interview with author.

246 *Sometimes the Germans offered . . . :* Frank Cerenzia, unpublished POW diaries, 1944–1945, hard copy.

246 *"We needed a radio . . .":* Tanner, interview with author.

246 *"Feb. 23, 1945, was a big day . . .":* Cerenzia, unpublished POW diaries.

246 *Patton defied General Eisenhower's . . . :* Alan Axelrod, *Patton's Drive: The Making of America's Greatest General* (Lanham, MD: Lyons Press, 2010), 140.

246 *"moving towards the Rhine . . .":* Axelrod, *Patton's Drive,* 140.

246 *"armored reconnaissance":* Major General Michael Reynolds, "Patton's End Run," Warfare History Network, November 6, 2018, https:// warfarehistorynetwork.com/daily/wwii/pattons-end-run.

247 *"swollen by the snow . . .":* Reynolds, "Patton's End Run."

247 *"Have taken Trier with . . .":* Reynolds, "Patton's End Run."

247 *"take the Rhine on the run . . .":* Eric Larrabee, *Commander in Chief: Franklin Delano Roosevelt, His Lieutenants, and Their War* (Annapolis, MD: Naval Institute Press, 2004), 492.

247 *"six battalions were over . . .":* Reynolds, "Patton's End Run."

247 *"For God's sake, tell the world . . .":* Patton quoted in Alexander McKee, *The Race for the Rhine Bridges, 1940, 1944, 1945* (London: Souvenir Press, 2001), Kindle edition.

247 *"without benefit of aerial bombing . . .":* Steven L. Ossad, *Omar Nelson Bradley: America's GI General, 1893–1981* (Columbia, MO: Univ. of Missouri Press, 2017), 322.

248 *"In the period from January 29 . . .":* Patton quoted in Earle Rice, *George S. Patton* (Langhorne, PA: Chelsea House, 2003), 103.

248 *"to take a piss in the Rhine . . .":* Reynolds, "Patton's End Run."

248 *"I have just pissed . . .":* William Kristol, "Men at War," *The Weekly Standard,* January 23, 2012.

251 *"a high point . . .":* Fox, *But You Made the Front Page!,* Kindle edition.

252 *"screaming and rifle . . ."*: Morse, *Sitting Duck Division*, Kindle edition.

253 *"given clearance to attack . . ."*: Frampton Jr., "Ich bin ein Kriegsgefangener."

253 *"We were lucky . . ."*: Fox, *But You Made the Front Page!*, Kindle edition.

254 *"We saw flashes of artillery . . ."*: Morse, *Sitting Duck Division*, Kindle edition.

255 *"and decided we had to . . ."*: Morse, *Sitting Duck Division*, Kindle edition.

255 *"After dark . . ."*: Kevin Lee Finneran, declassified confidential testimony of the War Crimes Office, Judge Advocate General's Department, War Department, sworn testimony, July 14, 1945 (NARA; Washington, DC: National Archives).

256 *"Sam let himself down . . ."*: Christian Wicks, declassified confidential testimony of the War Crimes Office, Judge Advocate General's Department, War Department, sworn testimony, July 14, 1945 (NARA; Washington, DC: National Archives).

257 *"An American colored soldier . . ."*: Frank Smith, declassified confidential testimony of the War Crimes Office, Judge Advocate General's Department, War Department, sworn testimony, July 14, 1945 (NARA; Washington, DC: National Archives).

258 *"an alarm went off . . ."*: James Lindow, declassified confidential testimony of the War Crimes Office, Judge Advocate General's Department, War Department, sworn testimony, August 1, 1945 (NARA; Washington, DC: National Archives).

258 *"Harris was shot, kicked . . ."*: Lindow, declassified confidential testimony.

258 *"A Negro Staff Sergeant . . ."*: James M. Hennessy, declassified confidential testimony of the War Crimes Office, Judge Advocate General's Department, War Department, sworn testimony, July 14, 1945 (NARA; Washington, DC: National Archives).

258 *"I saw bullet holes . . ."*: Elmer Kraske, classified testimony (since declassified) given during the investigation into the murder of Sgt. Sam Harris, conducted by the War Crimes Office, Judge Advocate General's Department, War Department, sworn testimony, October 1, 1945 (NARA; Washington, DC: National Archives).

259 *"The fury of the guards . . ."*: Peterson, *Healing the Child Warrior*, 52.

259 *"Two guards carried . . ."*: Morse, *Sitting Duck Division*, Kindle edition.

264 *"The danger was in the confusion . . ."*: Peterson, *Healing the Child Warrior*, 54.

264 *"We moved out in a rush . . ."*: Peterson, *Healing the Child Warrior*, 55.

264 *"I joined them, groaning . . ."*: Peterson, *Healing the Child Warrior*, 55.

266 *"We were told, 'Stay away . . ."*: Morse, *Sitting Duck Division*, Kindle edition.

268 *"We were mad . . ."*: Gunvalson, POW Oral History Project.

269 *"An American Sherman tank . . ."*: Friedman, interview with author.

271 *"The result was that although . . ."*: Fox, *But You Made the Front Page!*, Kindle edition.

272 *"torment my tormentors . . .":* Fox, *But You Made the Front Page!*, Kindle edition.

273 *"We arrived at about the same time . . .":* Frampton Jr., "Ich bin ein Kriegsgefangener."

274 *"You could have played any kind . . .":* Gunvalson, POW Oral History Project.

275 *"Foot soldiers have a one-sided . . .":* Peterson, *Healing the Child Warrior*, 54.

277 *"A glass of grapefruit juice . . .":* Stern, interview with author.

279 *"The washroom was the best . . .":* Cerenzia, unpublished POW diaries.

278 *"Although I was born . . .":* Fox, *But You Made the Front Page!*, Kindle edition.

281 *"Couldn't hit a mine for* sheet!*":* Morse, *Sitting Duck Division*, Kindle edition.

281 *"Cough, cough went the M1 . . .":* Morse, *Sitting Duck Division*, Kindle edition.

286 *"When we reached the coastline . . .":* Morse, *Sitting Duck Division*, Kindle edition.

286 *"We figured that we . . .":* Frampton Jr., "Ich bin ein Kriegsgefangener."

287 *"Paul, our apartment is . . .":* Stern, interview with author.

288 *"We sat there gorging . . .":* Stern, interview with author.

288 *"I went to Lake Placid . . .":* Tanner, interview with author.

289 *"The time was too short . . .":* Tanner, interview with author.

289 *"I'd finished college . . .":* Tanner, interview with author.

307 *"The lesson of that day . . .":* Lester Tanner, sworn affidavit in support of posthumous awarding Congressional Medal of Honor to Roddie Edmonds, June 10, 2014.

309 *"A 'Righteous' is a non-Jew who . . .":* "What Does 'Righteous Among the Nations' Mean? Irena Steinfeldt Explains," NBC News, January 27, 2016, https://www.nbcnews.com/video/what-does-righteous-among-the -nations-mean-irena-steinfeldt-explains-609661507787.

310 *"It's an attempt to recognize . . .":* "What Does 'Righteous Among the Nations' Mean?"

Selected References

Atkinson, Rick. *The Guns at Last Light: The War in Western Europe.* Vol. 3, 1944–1945. New York: Henry Holt, 2013.

Atkinson, Rick, with Kate Waters. *Battle of the Bulge,* the Young Readers Adaptation. New York: Henry Holt, 2015.

Beevor, Antony. *Ardennes 1944: The Battle of the Bulge.* New York: Penguin Books, 2015.

Berga: Soldiers of Another War. PBS documentary. Written and directed by Charles Guggenheim. Arlington, VA: Public Broadcasting System, 2002.

Caddick-Adams, Peter. *Snow and Steel: The Battle of the Bulge, 1944–45.* United Kingdom: Oxford University Press, 2015.

Churchill, Winston S. *Triumph and Tragedy.* Boston: Houghton Mifflin, 1953.

Cohen, Roger. *Soldiers and Slaves: American POWs Trapped by the Nazis' Final Gamble.* New York: Alfred A. Knopf, 2005.

Cole, Hugh M. *The Ardennes: The Battle of the Bulge.* United States Army in World War II, the European Theater of Operations. Washington, DC: Center of Military History United States Army, 1965.

Dupuy, R. Earnest. *St Vith: Lion in the Way: The 106th Infantry Division in World War II.* Washington, DC: Infantry Journal Press, 1949.

Footsteps of My Father. Documentary. Directed by Paul Allman. West Orange, NJ: Jewish Foundation for the Righteous, 2018.

Frankl, Viktor E. *Man's Search for Meaning.* Boston: Beacon Press, 1949.

Fussell, Paul. *The Boys' Crusade: The American Infantry in Northwestern Europe, 1944–1945.* New York: Modern Library, 2003.

George, Denise, and Robert Child. *The Lost Eleven: The Forgotten Story of Black American Soldiers Brutally Massacred in World War II.* New York: Dutton Caliber, 2017.

GI Jews: Jewish Americans in World War II. PBS documentary. Directed by Lisa Ades. Arlington, VA: Public Broadcasting System, 2018.

Hilberg, Raul. *The Destruction of the European Jews.* Chicago: Quadrangle Books, 1961.

Kershaw, Alex. *The Longest Winter: The Battle of the Bulge and the Epic Story of World War II's Most Decorated Platoon.* Cambridge, MA: Da Capo Press, 2005.

Kershaw, Ian. *The End: The Defiance and Destruction of Hitler's Germany, 1944–1945*. New York: Penguin, 2011.

King, Martin, Ken Johnson, and Michael Collins. *Warriors of the 106th: The Last Infantry Division of World War II*. Philadelphia and Oxford: Casemate Publishers, 2017.

MacDonald, Charles B. *Company Commander: The Classic Infantry Memoir of WWII*. United Kingdom: Burford Books, 1947.

MacDonald, Charles B. *A Time for Trumpets: The Untold Story of the Battle of the Bulge*. New York: Perennial, 1997.

McManus, John C. *Alamo in the Ardennes: The Untold Story of the American Soldiers Who Made the Defense of Bastogne Possible*. Hoboken, NJ: John Wiley & Sons, 2007.

McMillin, Woody. *In the Presence of Soldiers: The 2nd Army Maneuvers & Other World War II Activity in Tennessee*. Nashville, TN: Horton Heights Press, 2010.

Merriam, Robert E. *Dark December: The Full Account of the Battle of the Bulge*. Chicago: Ziff-Davis, 1947.

Messenger, Charles. *Sepp Dietrich: Hitler's Gladiator*. New York: Brassey's, 1988.

Morse, John. *Sitting Duck Division: Attacked from the Rear*. San Jose, New York, Lincoln, Shanghai: Writers Club Press, 2001.

Nobécourt, Jacques. *Hitler's Last Gamble: The Battle of the Ardennes*. Translated from the French by R. H. Barry. London: Chatto & Windus, 1967.

Paldiel, Mordecai. *The Paths of the Righteous: Gentile Rescuers of Jews During the Holocaust*. Hoboken, NJ: KTAV Publishing House, 1993.

Parker, Danny S. *Battle of the Bulge: Hitler's Ardennes Offensive, 1944–1945*. Cambridge, MA: Da Capo Press, 2004.

Peterson, Richard. *Healing the Child Warrior: A Search for Inner Peace*. Largo, FL: CombatVets Network, 1992.

Pyle, Ernie. *Brave Men*. New York: Henry Holt, 1944.

Reynolds, Michael. *Men of Steel: I SS Panzer Corps: The Ardennes and Eastern Front, 1944–45*. New York: Sarpedon, 1999.

Schrijvers, Peter. *Those Who Hold Bastogne: The True Story of the Soldiers and Civilians Who Fought in the Biggest Battle of the Bulge*. New Haven, CT, and London: Yale Univ. Press, 2014.

Sheaner, Herb. *Prisoner's Odyssey*. Bloomington, IN: Xlibris Corporation, 2009.

Strawson, John. *The Battle for the Ardennes*. London: B. T. Batsford Ltd., 1972.

Toland, John. *Adolf Hitler: The Definitive Biography*. New York: Anchor Books, 1976.

Toland, John. *Battle: The Story of the Bulge*. New York: Random House, 1959.

Whitlock, Flint. *Given Up for Dead: American GIs in the Nazi Concentration Camp at Berga*. New York: Westview Press, 2005.

Wiesel, Elie. *Night*. New York: Hill and Wang, 1960.

About the Authors

CHRIS EDMONDS is senior pastor of Piney Grove Baptist Church in Maryville, Tennessee, and chief executive officer of Roddie's Code, LLC, and the Roddie Edmonds Foundation, organizations committed to extending the legacy and leadership of Master Sergeant Edmonds to future generations. Pastor Edmonds also teaches leadership development to military leaders at the University of Alabama in Huntsville. He earned his bachelor's degree in business at the University of Tennessee and his master's degree in religion at Liberty Baptist Theological Seminary.

DOUGLAS CENTURY is the author and coauthor of numerous bestselling books, including *Hunting El Chapo*, *Under and Alone*, *Brotherhood of Warriors*, *Barney Ross: The Life of a Jewish Fighter*, and *Takedown: The Fall of the Last Mafia Empire*, a finalist for the 2003 Edgar Award for Best Fact Crime. Century is a contributing editor at *Tablet* magazine, and his journalism has appeared in the *New York Times*, *Rolling Stone*, *Billboard*, and *The Guardian*. Born and raised in Canada, he graduated from Princeton University and now lives in New York with his daughter.